Elements of Project Management

Plan, Schedule, and Control

Second Edition

M. Pete Spinner

 Prentice Hall, Englewood Cliffs, New Jersey 07632

Library of Congress Cataloging-in-Publication Data

Spinner, M.
    Elements of project management : plan, schedule, and control / M.
Pete Spinner. -- 2nd ed.
      p.   cm.
    Includes index.
    ISBN 0-13-253246-8
    1. Industrial project management.   I. Title.
HD69.P75S68 1992
658.4'04--dc20
                                            91-4004
                                              CIP

Editorial/production and interior
  design: *Carol L. Atkins*
Acquisition editor: *Michael Hays*
Freelance coordinator: *Brendan M. Stewart*
Editorial assistant: *Dana L. Mercure*
Cover designer: *Bruce Kenselaar*

Copy editor: *Zeiders & Associates*
Marketing manager: *Alicia Aurichio*
Prepress buyer: *Kelly Behr*
Manufacturing buyer: *Susan Brunke*

The publisher offers discounts on this book when ordered in bulk quantities. For more information, write: Special Sales/Professional Marketing, Prentice Hall, Professional & Technical Reference Division, Englewood Cliffs, NJ 07632

Printed in the United States of America
10 9 8 7 6 5 4 3 2 1

ISBN 0-13-253246-8

Prentice-Hall International (UK) Limited, *London*
Prentice-Hall of Australia Pty. Limited, *Sydney*
Prentice-Hall Canada, Inc., *Toronto*
Prentice-Hall Hispanoamericana, S.A., *Mexico*
Prentice-Hall of India Private Limited, *New Delhi*
Prentice-Hall of Japan, Inc., *Tokyo*
Simon & Schuster Asia Pte. Ltd., *Singapore*
Editora Prentice-Hall do Brasil, Ltda., *Rio de Janeiro*

This book is dedicated to that person whose unequaled zeal
for knowledge provided the inspiration and encouragement
for this endeavor

Contents

Preface to the Second Edition

Elements of Project Management was written to provide the basic fundamentals for arranging a plan and schedule for a project, as well as to provide techniques for monitoring and controlling the project once it is under way. The revisions in this edition reflect improvements in the fundamentals associated with project management that have occurred since the original edition was published.

The most significant developments that have occurred are (1) the phenomenal growth of the personal computer, permitting the concept of project management to expand into virtually every type of endeavor; and (2) the current emphasis in business and industry on participative management and employee involvement (PM/EI). As PM/EI is inherent in the application of so many project management principles, project management rules and techniques are now being incorporated into the standard operating procedures of many firms.

The original edition, with the changes that have been incorporated in the second edition, has been used by the author in training sessions and workshops to prepare project teams for planning, scheduling, and controlling major programs. In its complete and final form, this book will be useful in the following areas:

1. Universities and colleges with business and engineering curricula that recognize project management as a course requirement

2. Learning centers in business and industry that offer project management instruction

3. Intensive two- or three-day basic project management training or seminars conducted at continuing education centers

4. Self-training and reference material

The organization of chapters remains the same. Except for Chapter 7, *The Role of the Computer*, which has been revised completely to reflect the use of project management software for personal computers, each chapter continues to cover the earliest fundamentals, supplemented by up-to-date improvements and changes. A significant addition is the work breakdown structure (WBS), which has made it possible to bring organizational expedience into the planning process. PM/EI exponents will appreciate this discipline.

Chapter 6, *Resource Leveling*, has been updated by including not only computer-generated histograms, but an analysis that accompanies the charts. All of the sample projects used in this chapter and in other chapters are accompanied by revised planning diagrams that illustrate the importance of WBS for project organization.

Of interest to those involved in constructing planning diagrams is the arrow diagramming approach (i, j), used to better illustrate the significance of the WBS. Project members who make up a project team and who are responsible for specific groups of project items will recognize the utility of the arrow diagramming approach for this purpose.

In closing, I would like to acknowledge those persons whose help and encouragement have made this endeavor possible. I first recognize my wife, Margaret, who has remained patient and understanding while I have worked on earlier books but who, this time, in addition, has spent many hours typing the "tough stuff," assembling material, and doing all the unpopular jobs. Also, a very special thanks to Nan Nishanian, who among so many other duties, spent countless hours on the computer developing the type of computer-generated reports used in this book. I would also like to thank Kim Bondy, Metier Systems, for her contributions and suggestions, and for providing several special computer printouts.

A very special thanks goes to the suppliers of Pertmaster Advance, the project management package used extensively in this book. Specifically, I am in debt to Mr. Jim O'Hare, Vice-President, Projectronix, Los Altos, California, for his assistance; and to Mr. Charles Jordan, chairman and Managing Director, Pertmaster International, London, England, for both help and enthusiastic encouragement to continue this project.

M. Pete Spinner, P.E.
Southfield, Michigan

Preface to the First Edition

A few years after I began applying network planning with other project management principles in my work as an engineering and planning manager, I entertained the thought of documenting my experiences at an appropriate time. This book is my initial effort in realizing that goal. Supplementing my experiences are the research and lecture material that I use for teaching project management classes.

The book is written so that readers with varying educational backgrounds and experience can understand its contents. In addition to the fundamentals, the text expands on project management methods that will be useful to those with previous training and experience.

The readers, who may be our future planners, are shown the principles (or tools) used to arrange an effective plan and schedule, and after a project has started, the methods useful in controlling the direction of the project to its successful completion. All of these phases are presented in the sequence of how an actual project evolves, and a sample project is used as an illustration for applying project management techniques discussed throughout.

The book is arranged in such a manner as to illustrate first, the basic fundamentals of organized planning and scheduling; second, monitoring and controlling projects (with emphasis on such communication styles as status reports); third, handling project costs and labor allocation; and finally, computer application. Two chapters are devoted to a companion network planning method, Project Evaluation and Review Technique (PERT). This approach to project management implementation may be beneficial in some areas of business and industry, and of special note are the time and cost status reports illustrated in these sections.

When calculations are required, elementary arithmetic supplemented with simple graphs is used. Purposely avoided are any detailed analytical exercises that may be used to substantiate the derivation of several of these techniques. Using successful applications as examples can show more than any management science displays, such as linear programming or mathematical modeling.

The text emphasizes that planning and scheduling can be done even without a computer but that some applications can make good use of a computer. One chapter explains how the speed and accuracy of computers is especially useful for large, complex programs.

Included in the book are problems concerning major subjects that can be used for readers who want to determine how well they understand the elements of project management. Several sample projects are included that can be selected for a term project problem. These problems are designed so as to incorporate all aspects of project management covered in this book. Completed project problems are useful as reference material.

Most of the actual applications used are from my experiences at Ford Motor Company. This is where it all started for me in applying and developing the techniques explained in the book. I wish to acknowledge Ford Motor Company for their permission to use portions of a Ford training manual entitled, "Training Course 3315—Critical Path Method." Ford Motor Company is not however, responsible for the accuracy or content of any of the material included in this book; the contents are of my own design and I assume full responsibility for them.

I wish to express my appreciation to those who shared in the preparation of this book: Chairman Wayne H. Buell, President Richard E. Marburger, and Dean Richard E. Michel of the Lawrence Institute of Technology, for their support and cooperation; my son, David, and my cousin, Nina Mayers, for their proofreading efforts; Nan Scullin for her untiring efforts in preparing the initial diagrams; and a very special thanks to Louanne Snyder, my "Girl Friday" on this project, for her valuable suggestions and particularly for having the patience and ability to translate my handwriting into a typed manuscript.

M. Pete Spinner, P.E.
Southfield, Michigan

1
Introduction

A knowledgeable authority in project management has stated: "The requirements for project management are simple—you only need infinite patience, understanding, and wisdom."[1]

There is much truth in this statement, especially in the handling of modern-day projects that are complex and are of large proportions. However, there are tools available, simple to learn and apply, that can alleviate the problems encountered in project management. Among these tools are the following:

1. Planning, scheduling, and controlling time and costs.
2. Program reporting and forecasting time duration.
3. Cost reporting and forecasting total expenditures.
4. Use of computers, in conjunction with the above, especially for large, complex projects.

We shall list several definitions before proceeding with a discussion of the major topics.

[1] Eric Jennett, "Guidelines for Successful Project Management," *Chemical Engineering,* July 9, 1973.

Project management in the business and industry fields is defined as managing and directing time, material, personnel, and costs to complete a particular project in an orderly, economical manner; and to meet established objectives in time, dollars, and technical results.

One can define a *project* by means of the following distinguishing characteristics:

1. There is a specific start and a specific end point.
2. There is a well-defined objective.
3. The endeavor is unique and not repetitious.
4. The project usually contains costs and time schedules to produce a specified product or result.
5. A project cuts across many organizational and functional lines.

Project management principles are disciplines employed in planning, scheduling, and controlling a project. The most popular, *network planning,* is a relatively new technique used to help accomplish the successful practice of project management. There are other techniques that complement the application *network planning* and are included in the discussion in this book.

- Management by objectives
- Management by exception
- Cost analysis
- Personnel/labor allocation
- Participative management/employee involvement

The use of these principles is associated with careful, detailed planning; therefore, the user is forced to think through the project.

NETWORK PLANNING

The most widely used project management principle is network planning. This technique is used to plan, schedule, and control a project consisting of a group of interrelated jobs (which may also be called work items or activities) directed toward a common goal. The network planning method is especially useful for those projects that have a well-defined starting point and a well-defined objective; project performance is usually very good when using this method. Modified versions of the usual network planning methods are required to plan and schedule production control or process control operations or any type of activity that involves continual scheduling on a continuous flow of activities.

History of Network Planning

The initial undertaking in planning a project is the development of a graphical diagram. The idea of using diagrams to plot the progress of a project is quite old. Things that look like network diagrams appear in the literature as early as 1850. In particular, George Boole, who worked in the field of logic and algebra, used diagrams to explain propositions in logic and the flow of logical problems. The Prussians in the late nineteenth century developed diagrams to show tactical movements on battlefields. They showed where their troops would be, where the enemy troops would be, and how a battle would progress. One of the favorite pastimes of military people is still the construction of diagrams to show how famous battles occurred.

Another use of diagrams arose from the work of economists. An article published in 1944 indicates that economists developed arrow diagrams designed to show the flow of systems and the interrelation between systems. The diagrams look surprisingly like those used in network planning or some of the other systems that we have today.

The need for improved planning and progress evaluation techniques to help control the utilization of manpower, material, and facilities became apparent at approximately the same time in the 1950s. The pioneering application of the network diagram and the *critical path concept* was a jointly sponsored venture of E. I du Pont de Nemours and Company and the Sperry-Rand Corporation. The objective of this venture was to improve the planning, scheduling, and coordination of du Pont's engineering projects. By September 1957 an actual application was conducted on a pilot system using the UNIVAC I computer, and from this initial effort, network planning and the critical path method evolved.

Simultaneously, the Navy was devising a system to plan and coordinate the work of nearly 3,000 contractors and agencies on the Polaris missile program. From the Polaris project came the *Program Evaluation and Review Technique* (PERT), which is credited with helping advance Polaris development by at least two years.

Today many industries use networks to plan projects of varying size and complexity. Procurement and installation of tooling, building facilities, and machinery are being scheduled under network control. Network planning has many other applications. It can be used to plan design operations, construction projects, administrative programs, maintenance operations, model change-overs, and practically any other series of actions that, when combined, form a complete program having a start and a finish.

The multitude of names applied to network diagramming methods, such as CPM (Critical Path Method), PERT (Project Evaluation and Review Technique), PEP (Project Evaluation Procedure), and LESS (Least Cost Estimating and Scheduling), merely distinguish the application techniques. The most commonly used are CPM and PERT. The main distinguishing characteristic is that CPM traditionally has been considered activity-oriented, whereas PERT is event-oriented. However, these differences are gradually melting away, and the term *network planning* is now commonly used as a general term to describe all these types of operational programming.

The network diagramming procedure used in this text is the *Critical Path Method* (CPM), which requires that all jobs be completed with no allowance for failure. This can limit the use of CPM in such programs as planning research projects, feasibility studies, test programs, and preparation of proposals. Such projects will have situations where there are alternative approaches, and all of the paths may not only have an unpredictable outcome, but all but one may be aborted.

There is a network diagram system that allows for jobs to be started when there is uncertainty concerning completion of proceding jobs. The reader is advised to gain a most thorough knowledge of the CPM technique before attempting application of another network diagram system.

In network planning, the terms *planning* and *scheduling* are not synonymous. By definition, a *plan* is a proposed method of action or procedure. Planning may or may not be dependent on timing. Planning indicates *what* activities are to be accomplished. Scheduling is the development of a timetable that puts time estimates next to the plan and indicates *when* activities are to be accomplished.

Bar Charts

The traditional approach to planning and scheduling has been through the use of *bar charts,* which portray the timing and duration of each activity in a project. These charts are familiar to almost everyone in industry. Basically, they depict graphically the jobs to be done and the timing for each job, as shown in Figure 1-1. This type of bar graph indicates the beginning and end dates for each of Jobs A, B, C, and D. These jobs represent the engineering, material procurement, construction, equipment delivery and installation phases, respectively, of an engineering project. In many cases, we might find that one or more of these jobs is subdivided and detailed on a second chart. To illustrate, Job A might be divided into several major components, each of which would become a bar or a bar graph.

A chart of this kind provides some valuable information, but essential data are missing. From the standpoint of effectively planning, scheduling, and controlling a project, additional and more accurate information is required. For example, the relationships among engineering, procurement, construction, and equipment delivery and installation (activities represented by Jobs A, B, C, and D) cannot be shown and the following questions cannot be answered:

1. What parts of these jobs can be performed concurrently?
2. What parts of each job are necessary to be completed before other parts begin?
3. Must certain jobs or parts of jobs be given priority in order not to hold up completion of the project on schedule?
4. Do some jobs or parts of jobs have optional starting and end dates, and what specifically are these optional dates?

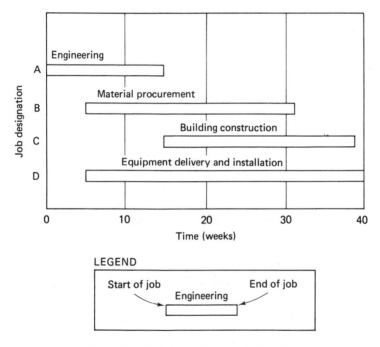

Figure 1-1 Basic type of bar graph (chart).

Henry L. Gantt, one of the pioneers in scientific management, developed many kinds of charts and records for planning purposes. About 40 years ago, Gantt developed connected bar charts which bear some resemblance to arrow diagrams. Different kinds of shading have been incorporated in bar graphs to show the activities that could be started early; however, there are limitations. Revisions to these charts are difficult to make and require a great deal of time to update, especially as projects have grown in size and complexity.

Using basically the same project elements as shown for the bar graph, a simplified *network diagram* can be illustrated graphically, as shown in Figure 1-2. Network diagrams overcome the deficiencies of bar chart construction by providing essential information necessary in planning projects:

1. Network diagrams explicitly show interrelations between jobs.
2. A network diagram shows which jobs can be done concurrently, which ones precede, and which ones follow other jobs.
3. Jobs with critical schedules are specified with their required beginning and completion dates.
4. Jobs of a noncritical nature are also shown with optional beginning and end dates.

Although bar charts have limitations for planning purposes, they have been refined over the years to provide an excellent communication expedient to management by summarizing the status of projects.

Network Planning Procedure

Planning, scheduling, and controlling projects, the three phases in the project management cycle, are handled separately for more effective results in the network planning procedure. The planning phase can be the most time consuming of the total project process: however, the time spent planning can also be the most rewarding. Planning is determining *what* work is to be done. *Planning* a project will follow these steps:

Step 1. *Establish Objectives*
a. State objectives that will be derived from the requirements which motivated the project.
b. List interim objectives or milestones that are significant in meeting the main objectives.

Step 2. *Develop a Plan*
a. List the jobs (or activities) that have to be done to complete the project.
b. Delineate the jobs by the following procedure: After the jobs have been decided upon, the relations between them should be determined. This involves a careful analysis of each job, including such items as:
(1) Which jobs precede and succeed other jobs on the list.
(2) Which jobs can be accomplished concurrently.
c. Portray the sequence in a network (arrow) diagram.

Determining what is to be done must be the result of a careful analysis of the project by those knowledgeable in the particular field. The network plan

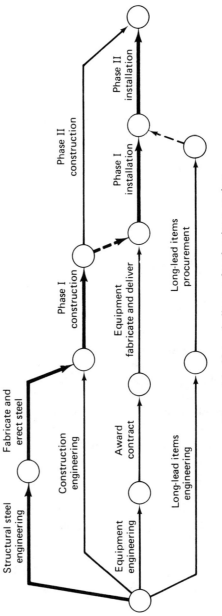

Figure 1-2 Network diagram for planning a project.

Structural steel engineering

Fabricate and erect steel

Construction engineering

Phase II construction

Phase I construction

Phase II installation

Phase I installation

Equipment engineering

Award contract

Equipment fabricate and deliver

Long-lead items procurement

Long-lead items engineering

7

then makes use of an arrow diagram to display graphically the sequence and interrelations of the jobs required.

After the network diagram has been completed, scheduling begins. The scheduling phase introduces the timing aspects of the job. Scheduling is determining *when* the work is to be done.

Scheduling a project will follow these steps:

Step 1. After the sequence of jobs has been planned and laid out in a network diagram, the timing can be established.
a. Estimate the time required to complete each of the jobs in the project.
b. Calculate the schedule.
c. Compute the available time to complete each job.
d. Identify the critical jobs.
e. Determine the float times of the noncritical jobs.

Step 2. If the project duration time that is calculated initially is not acceptable, make adjustments to the plan so as to meet a project deadline that is acceptable.

Step 3. Establish a calendar schedule. (Using a bar chart may portray the schedule effectively.)

Effective project control involves constant monitoring of each job in the project. Actual job progress is noted on charts or various other status reporting methods. Summary reports to management are submitted at periodic intervals, and these reports can be prepared weekly, biweekly, or monthly, depending upon the dynamics and scope of the project. The reports will project completion status (generally including costs) and will highlight critical items that may jeopardize the schedule.

Supplementing the network planning technique as the major tool in managing a project to a successful completion are such techniques as management by objectives; management by exception; cost analysis, including time/cost trade-offs (or cost minimizing); and resource allocation. Next, we define these techniques as they apply to project management.

MANAGEMENT BY OBJECTIVES

Once the goals have been identified, an orderly procedure can be set up so that all of the combined efforts are directed in such a way that the goals are achieved. Through network planning, application of *management by objectives* can readily be achieved:

1. Because you have a goal, you know whether or not you are on the right road.

2. You can assess results all along the course of the project.
3. By regularly assessing performance of your goals, you are on top of the project.
4. You will perform with maximum effectiveness by knowing what goals the project requires and how well you are performing in relation to these goals.

MANAGEMENT BY EXCEPTION
(RECOGNIZING PROBLEM AREAS)

Network planning facilitates application of the management by exception principle by readily identifying the critical operations. The critical operations usually make up about 20% of the project activities that can affect the overall progress. Through network planning there is a clear definition of just how far each of the other jobs can slip behind schedule without affecting overall progress. This permits true *management by exception,* since management can concentrate on the critical jobs. At the same time, limits are set up for the remaining jobs that must be met if the project is to be completed on schedule. Any slippage beyond these limits immediately signals the need for management attention.

COST ANALYSIS

Network planning also makes it possible to consider costs in the same way that project timing aspects are handled. Cost scheduling is helpful to those funding the project. It is a valuable tool for projecting cash flows and for preparing a planned expenditure distribution. On large projects cost schedules prepared in a disciplined manner are the basis of determining property taxes and depreciation schedules.

There are cost disciplines designed to ensure that project spending is contained within approved or authorized amounts. One effective project cost control technique used is the *indicated cost outcome.* It provides a periodic review and evaluation of the spending status of open projects to determine if spending is in line with approved authorizations. Specifically, an indicated cost outcome procedure will provide an early-warning system for potential project overspending or underspending. Planning future spending levels and comparing them with authorized amounts will determine whether anticipated project spending may require additional authorization.

The results of the planning phase should produce the least costly schedule for a feasible project duration. The *minimum-cost expediting technique* available through the use of the network planning procedure does just that. It produces schedules for a number of different project durations, each of these being the minimum-cost schedule for that particular project duration.

How can the project be rescheduled to meet the required deadline at minimum additional cost? The first step in shortening the duration of a project is to select the critical jobs that can be reduced. We will assume that unreasonable time estimates have already been identified and eliminated. A reduction in time can usually be made through expediting overtime, assignment of additional people, use of air freight, and so on, but at additional cost to the project.

Next, the additional cost involved in reducing project duration is matched with the benefits that can be realized when the time of the project is reduced. The objective is to reduce the duration of those jobs that can be shortened with least additional cost.

For this purpose, a cost-minimizing program has been developed as a management discipline. Through this discipline, rescheduling, including the selection of jobs to be accelerated, is accomplished in such a way that whether the final cost is smaller, larger, or unchanged, the added cost has been minimized.

LABOR ALLOCATION/LEVELING

In the initial phases of network analysis, there is no restriction on the availability of labor. However, an important problem in labor allocation is that only so many persons are available to perform the work in a project. Another problem in personnel scheduling is to make efficient use of the labor that is available. It is necessary to avoid labor-usage curves that show sharp peaks and valleys for a number of reasons:

1. Efficient use of labor will cut costs.
2. Effective use of labor will reduce unnecessary overtime.
3. A reputation for offering life-of-the-project employment will make it easier to attract and hold the best skilled labor.

All sorts of intuitive methods for leveling the use of workers have been tried. The intuition that enabled the practitioners of these methods to enjoy partial success also told them that purely intuitive methods were woefully inadequate. When intuitive methods are supported by a disciplined approach to resource allocation, quite satisfactory results can be obtained.

PARTICIPATIVE MANAGEMENT/EMPLOYEE INVOLVEMENT

While the practice of *participative management and employee involvement* (PM/EI) in business and industry is relatively new, its essential characteristics are inherent in applying project management techniques.

- *Participative management* (PM) can be defined as the combination of techniques and skills that you use to provide employees at all levels with opportunities to participate actively in key managerial processes affecting job-related matters.
- *Employee involvement* (EI) can be defined as the process by which employees at all levels have the opportunity to participate actively in key managerial processes affecting job-related matters.

Groups associated with PM and EI have closely related responsibilities and participate in many facets of operations: planning, organizing (analogous to developing the work breakdown structure), goal setting (allied to managing by objectives), and decision making. The overall goal of PM/EI is to develop a well-rounded management style that emphasizes a participative cooperative relationship in accomplishing work objectives.

Introducing PM/EI into business and industry may be characterized by the following guiding principle, included in the *Statement of Mission, Values and Guiding Principles* submitted by Ford Motor Company's board chairman: *"Employee involvement is our way of life.* We must treat each other with trust and respect."

APPLICATIONS IN BUSINESS AND INDUSTRY

The use of network analysis with the associated project management principles for project administration has become popular in industry, business, and government. The use of project management principles is growing because they aid in satisfying a need to assist in resolving the problems encountered from the ever-increasing complexities in modern-day projects.

Construction Industry

Because the work conditions that make up a construction project are dynamic in nature and changes are commonplace, network planning has become very popular in this industry. Arrow diagramming has become standard practice for planning and scheduling construction projects.

Changing conditions, such as fluctuating labor and design changes, are the rule rather than the exception in a typical construction project. Where alternative plans are constantly being evaluated, project management principles are the tools used to complete projects successfully. Rigid disciplines are needed to keep programs on schedule within budgeted amounts.

Industrial Projects

As industrial expansion is closely allied to construction operations, the use of network planning in all facets of business activity, including industrial expansion, has been most productive. Marketing objectives, financing methods, facility evaluations, and so on, are laid out with alternative plans and made available for review by management at all levels. The logic and timing developed from arrow diagrams serve as reliable supporting data in feasibility studies and business planning presentations.

Marketing Programs

Planning a marketing program can involve the efforts of many diverse individuals in research and development, engineering, manufacturing operations, sales, and various management groups. These personnel, all of whom have an interest in the program, are essential to market a product successfully. A marketing program may begin with an extensive market analysis survey of potential consumers and continue through the construction of a new manufacturing plant, possibly in a foreign country. The inclusion of an advertising campaign adds to the complexity of the project. To plan such a program requires a great deal of coordination of the participating activities.

In recent years, marketing groups have adopted the network analysis approach in planning marketing programs. A network plan will clearly show all the essential steps in conducting the program, and it will also define the responsibilities of the participating groups. The coordinating requirements are also made visible, so that no group can overlook its relationship to others.

As this important business segment involves a spectrum of diverse activities, the first sample problem used in the book is a marketing project. The project, entitled "A New Product Introduction" illustrates many project management techniques. As the sample problem will illustrate, the marketing project need not be long or very detailed to be effective. What is important is the ability of the method to provide an adequate plan, which the arrow diagramming approach accomplishes for this particular program.

Hospital Capital Programs

Delays so commonly experienced in hospital projects have been avoided by network planning, which is diagrammed from the beginning, the predesign phase, to the end, the occupancy phase. The technique has helped to create positive management interest early in the program. By recognizing problems early, hospital officials have avoided months of "drifting," so prevalent in hospital programs that utilize unrealistic planning methods.

Government Activities

For almost 20 years network planning diagrams have been included in the specifications prepared for proposals submitted for governmental activities, including the U.S. Navy, U.S. Army, U.S. Air Force, Department of Defense, and National Aeronautic and Space Agency. The major reason for using network diagramming is that governmental agencies require so much service and approval that documentation of their activities is a necessity. Network diagramming lends itself to these demands and its utility has extended beyond this purpose to allow it to be used in other areas. In addition to its use in construction projects and planning programs, network planning diagrams are being used for administrative plans.

Computer Application

Using a computer to apply the technique of network analysis for large projects is essential. A computer saves substantial time in determining which jobs must be given priority (those on the critical path), in making schedule changes, in leveling the use of workers, and in accomplishing a variety of other planning, scheduling, and control functions. Chapter 7 is devoted entirely to use of the computer in project management applications.

ADVANTAGES IN APPLYING PROJECT MANAGEMENT PRINCIPLES

Today's planner can use project management techniques to guide and control the course of a project with more confidence than by use of any of the older methods. However, these techniques cannot and will not replace the value judgments of human beings. The results are only as good as the effort that is put into them.

To be a useful tool, a network must reflect reality as closely as possible and be continuously revised to keep it accurate. Above all, for the utmost effectiveness, complete backing by top management is essential. They must understand and accept the method if it is to be used successfully. With effort, participation, and cooperation, the advantages are many. Following are some of the more important benefits that will be gained by its use:

1. It makes participants think a project through in greater detail.
2. It enables more efficient use of resources, such as personnel, equipment, space, and money.
3. It gives a clear picture of the project that is readily communicable to everyone involved, including new personnel.
4. It enables true management by exception. Management can react quickly to the critical items that may jeopardize a project.
5. It enables real control of projects that previously may have been too unwieldy for anyone to understand.

6. It provides management with data on which to base plans for minimizing investment costs or maximizing return on investment.

7. It enables quick rescheduling of a project to meet changing or unpredictable conditions.

8. It can help improve labor relations.

9. It is a proven system for planning, scheduling, and controlling projects.

10. It provides a graphic picture of the work by showing the proper relationships among the project work items.

2
CPM/Project Planning

Planning is a vital management function; in fact, most management authorities consider planning to be the most important function. Yet the responsibility of developing plans is often superseded by such items as troubleshooting or reviewing ongoing jobs or operations. Although these tasks are important, they cannot replace the time that needs to be allotted for the planning phase.

The quality of planning is also an important consideration in the successful outcome of a project. After experiencing a trying project, one manager surmised: "Possibly the reason that the plan failed was that there may not have been a plan at all." Providing a good technique for planning a project is the purpose of this chapter.

There are three phases in the project management cycle: planning, scheduling, and control. In planning, we determine *what* has to be done in accomplishing a project, establishing the sequence of work, and specifying the interrelations between jobs. Timing, or *when* the work is to be done, is not generally considered during the planning phase. In more detail, the planning phase follows this procedural outline:

Step 1. *Establish Objectives*
a. Objectives are normally established by higher management and/or dictated by a particular company goal.
b. Interim objectives or milestones may also be specified.

Step 2. *Develop a Plan*
a. Prepare a list of jobs required for the project.
b. Determine the relationships between the jobs:
    (1) What jobs precede a given job on the list?
    (2) What jobs follow a given job on the list?
    (3) What jobs can be accomplished simultaneously?
c. Portray these relationships in an arrow diagram.

In the next phase, scheduling, we are concerned with the timing aspects, that is, how much time each job is expected to require for completion, and when each job will be scheduled to begin and end. People familiar with the work required must provide the information as to what is to be done, the sequence to be followed, and the time required to complete each job.

For complex projects, the computer is used to make the critical path timing calculations accurately and rapidly. The use of the computer is most helpful in the third phase, project control, for providing timely follow-up information in a convenient and effective form, and to update and adjust the scheduling data quickly as the job progresses. Providing reports on the current status of the project is an important communication link.

To summarize, planning, scheduling, and control of a project will follow this sequence:

1. *Project Planning*
    a. Objectives
    b. Content of project
    c. Arrow diagram
2. *Project Scheduling*
    a. Time estimates
    b. Timing calculations
    c. Job scheduling
3. *Project Control*
    a. Follow-up
    b. Updating
    c. Reporting

Drawing the arrow diagram is a part of the planning phase, and should be done initially without reference to the timing aspects. Separating planning and scheduling helps to simplify their accomplishments and should result in a more effective job being done in both areas.

THE ARROW DIAGRAM

As the final part of the planning phase, the arrow diagram shows the sequence of jobs to be accomplished and their interrelations. The arrow diagram provides several important advantages:

- A disciplined basis for planning a project
- A clear picture of the scope of the project that can be easily read and understood by someone who is not familiar with the project
- A means of communicating what is to be done in the project to other people involved, some of whom are not familiar with its details
- A vehicle for use in evaluating alternative strategies and objectives
- A means of defining interconnections among jobs and pinpointing how the responsibilities for accomplishing the jobs fit into the total project
- Assistance in refining the design of the project
- An excellent vehicle for training project personnel

After project personnel have analyzed the project carefully, determined what jobs must be performed, and outlined the flow of work to be followed, they translate the results into an arrow diagram. An arrow is drawn for each job. The sequence of the arrows indicates the flow of work from the beginning of the project to the end. The diagram has one beginning point and one end point. Arrow junctions are called *nodes*. They are usually numbered, which is useful for identifying the location of arrows in the network.

Arrows

In an arrow diagram, an arrow is used to represent a *job* or *activity*.

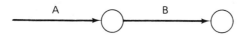

- The work is assumed to flow in the direction in which the arrow points.
- A job therefore begins at the left end of the arrow and finishes at the right end (at the head of the arrow).
- The arrows are not time-scaled.
- Job sequences are indicated by the way the arrows are interconnected. In the following illustrations, the arrow interconnections show job sequences:

Job A must be completed before Job B is started.

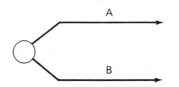

Job A can be done concurrently with Job B.

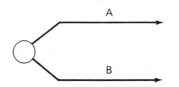

Jobs A and B can be done concurrently and must be completed before Job C begins.

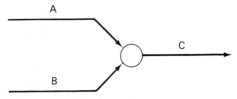

Two types of arrows are used in diagramming:

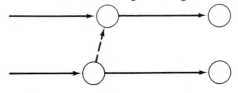

1. Solid arrows represent jobs or activities.
2. Dashed or dummy arrows show special interrelations.

The solid arrow has several features:

1. It represents a job or activity that consumes time (has a duration) and resources.
2. Work is assumed to flow in the direction in which the arrow points.
3. The length can vary (time or duration is not indicated by the length).

There is no simple answer to the question: How much work should be grouped under a job? A job may represent work of considerable complexity or a small detail of the total project, depending on the purpose of the network.

In a network designed for management use, each arrow would represent a major activity of the project. The diagram would include only major elements, thus providing an overall view of the entire project. On the other hand, if the network will be used by persons who will supervise or perform the work, each arrow should represent a small segment of the project. Often, both a general diagram, consisting of major elements only, and a detailed diagram will be prepared for a project.

Nodes

In network diagramming, the beginning and end of each arrow is called a *node* or an *event*. A node or event is a point in time and has no duration. For convenience of notation, an arrow is sometimes said to go from i to j. The i represents the beginning of the arrow; the j represents the end of the arrow. An arrow is uniquely identified by its i and j; for example, "Arrow 2,3" is the "B" activity in the following illustration:

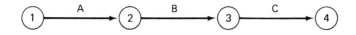

- The circles with numbers 1,2,3, and 4 are nodes.
- Each arrow (job or activity) is identified by two nodes—the *i* node at the beginning of the arrow and the *j* node at the end.
- Job A is identified as Job 1,2; Job B as Job 2,3; and so on.

Work Sequence

The arrow diagram must indicate the *sequence* in which the work will be performed according to a plan. Since some variation in work sequence in a project is often possible, it should be noted that every arrow diagram represents someone's specific plan for accomplishing the project. As the plan changes, so must the diagram.

Three questions are used as guides in arranging the sequence for each job:

1. What precedes the job?
2. What follows it?
3. Which job(s) can be performed concurrently with it?

A review of some of the basic points involved in developing an effective diagram is provided later in the chapter in the sample problem, "A New Product Introduction."

Effective planning of the jobs or activities in a project requires that certain techniques need to be adhered to in diagramming the network. Several of the points to be considered when diagramming are shown in the following illustration:

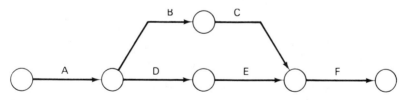

- The position of Job B in the illustration:
  a. Job B follows Job A.
  b. Job B precedes Job C.
  c. Job B can be performed concurrently with Jobs D and E.
- Jobs C and E must be completed before Job F can begin.
- Jobs B and D may start as soon as Job A is completed.
- When two or more arrows begin at the same node, this does not mean that the work they represent will be scheduled to start at the same time. For

example, Jobs B and D may start as soon as Job A is completed; however,
their scheduled starting times depend on how much "float" each job has.

- Two or more arrows that end at the same node do not necessarily have an
  identical finish date. Jobs C and E must both be completed before Job F
  can begin, but one might finish before the other

- If Job H can start after half of Job G is completed, Job G should be
  divided into two parts: G1 and G2.

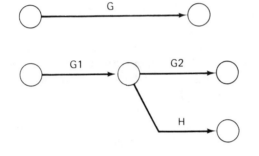

- PERMITTED

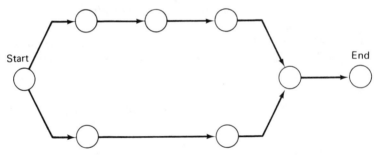

- NOT PERMITTED
  (Looping)

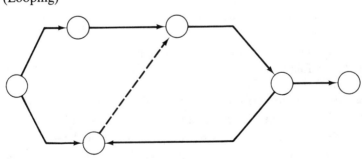

- NOT PERMITTED
  (Two Starting Nodes)

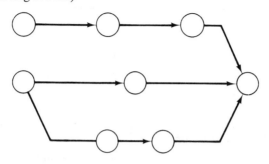

Dummy Arrows

In an arrow diagram, *dummy arrows* express relations between jobs that are not indicated by solid arrows. The dummy arrows are used as restraints and as a convenience in drawing the network. A dummy arrow can be used to maintain the node identification.

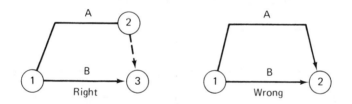

Dummy arrows are drawn as dashed lines and, if possible, are placed at the end of an activity. Dummy arrows do not represent a job or activity; they have no duration and do not consume resources, such as labor, dollars, and so on. The principal uses of dummy arrows can be illustrated as follows:

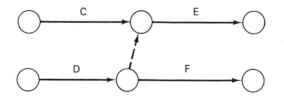

1. • Jobs C and D must be completed before Job E can begin.
   • Job F can be started as soon as Job D is completed.

The following diagram would not properly represent these relations:

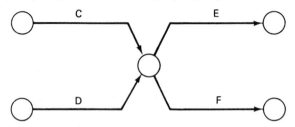

- Job F is overstrained. This diagram shows that F must follow C and D, whereas F is required only to follow D.

2. The dummy arrow may also be used for convenience or to save time in drawing the diagram. To illustrate, assume that a diagram has been drawn which has over 100 activities. An activity (Job M) occurs near the bottom of the network and must be followed by an activity (Job J) that occurs near the top of the network. Instead of redrawing the network so that Jobs M and J could be shown with solid arrows, which would require considerable extra work, a dummy arrow is drawn from Job M to Job J.

3. Another use of the dummy arrow is to identify arrows uniquely—to avoid duplicating a set of node numbers when identifying different arrows. This situation is illustrated in the following diagrams:

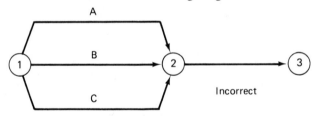

Since "Job 1,2" could refer to any one of three jobs, ambiguous identification results. This situation is called *branching*. Computer programs, which use node numbers for identification, would not be able to process the network data properly. In the following diagram each arrow has a unique node number:

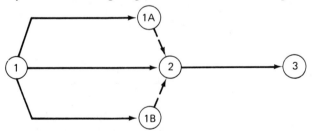

In general, nodes should be numbered in such a way that each job has a unique identification and can be located easily in the network. Beyond this, each computer program for processing project management networks may have its own special node number requirements. Only a few typical requirements will be discussed here.

Some older computer programs require that nodes be numbered consecutively in ascending order; the integers "1, 2, 3, . . . " are used and each arrow flows from a lesser integer to a greater integer. However, very few current computer programs require consecutive numbers. Some current programs still require that arrows be numbered with ascending integers, but they do not require that consecutive numbers be used. Consecutive numbering could cause a problem when the network is subject to continual updating, requiring new arrows to be added and some existing ones to be eliminated. To avoid this, a numbering sequence with gaps between integers is often used: for example, "1, 5, 10, 15,. . . . " As arrows are added later, unused node numbers may be assigned from the proper interval to maintain an ascending sequence.

More recent programs contain sequencing routines within them, thus eliminating the ascending-order requirements. These programs generally will accept either numeric or alphabetic node designations or both. It should be noted, however, that it is easier to follow the flow of work visually through the diagram and to locate a specific activity if a general ordering scheme is employed.

The person numbering a diagram should inquire as to whether the computer program to be used subsequently contains any numbering restrictions. The new product introduction project will be used to illustrate the planning process.

DEVELOPMENT OF A NETWORK DIAGRAM: A NEW PRODUCT INTRODUCTION

An established company has decided to add a new product to its line. It will buy the product from a manufacturing concern, package it, and sell it to a number of distributors selected on a geographical basis. Market research has indicated the volume expected and the size of sales force required. The company wants to place the product on the market as soon as possible.

1. Establish objectives. Top management has directed that the company's new product be introduced as rapidly as possible. The project manager, more specifically, is committing the project team to completing the project by October 19, 1991, contingent upon receiving approval to start the project April 1, 1991. (The completion date is also contingent upon the availability of resources.)

2. Establish the jobs required. The first step in developing the diagram is to establish the jobs required for accomplishing the project. Generally, the list of job descriptions is developed by personnel experienced in the type of work involved (see Figure 2-1).

Proper listing of every step required to accomplish the project is of great importance. It is vital that all phases of work be encompassed by the jobs listed. Any omissions will cause inaccuracies in scheduling and may result in failure to complete the project on time.

A job description list should be made which lists the major phases of work involved in a project. After the jobs have been decided upon, the relations between them should be determined. This involves an analysis of each job.

There are several basic tasks that must be performed during the development of an arrow diagram. The procedure followed in accomplishing these tasks may vary from person to person. For example, one diagrammer may obtain the list of jobs for the project and then jot down the relations between jobs before beginning the network; another person might go from the job list right into the diagramming and develop the job relations as he or she proceeds through a series of rough and final diagrams.

The basic tasks to be performed include listing the tasks required to complete the project, dividing the tasks into groups representing major sections of the project, determining the relations between the tasks, and drawing the arrow diagram.

ACTIVITIES IN THE PROJECT

The following activities are to be planned along with the relations among them:

1. Organize the sales office: Hire the sales manager.

2. Hire sales personnel: The sales manager will recruit and hire the salespeople needed.

3. Train sales personnel: Train the salespeople hired to sell the product to the distributors.

4. Select advertising agency: The sales manager will select the agency best suited to promote the new product.

5. Plan advertising campaign: The sales office and the advertising agency will jointly plan the advertising campaign to introduce the product to the public.

6. Conduct advertising campaign: The advertising agency will conduct a "watch for" campaign for potential customers.

7. Design package: Design the package most likely to "sell".

8. Set up packaging facility: Prepare to package the products when they are received from the manufacturer.

9. Package initial stocks: Package stocks received from the manufacturer.

10. Order stock from manufacturer: Order the stock needed from the manufacturer. The time given includes the lead time for delivery.

11. Select distributors: The sales manager will select the distributors whom the salespeople will contact to make sales.

12. Sell to distributors: Take orders from the distributors for the new product, with delivery promised for the introduction date. If orders exceed stock, assign stock on a quota basis.

13. Ship stock to distributors: Ship the packaged stock to the distributors in accord with their orders or quota.

Figure 2-1 Activities required for "A New Product Introduction."

3. Divide the jobs into groups representing major sections of work.

As an initial aid to establishing the relations between jobs, the project may be divided into groups of closely related functions under the categories of work to be performed. In the example, the jobs have been grouped under the major headings "Stock," "Packaging," "Sales," "Distributors," and "Advertising." These logical groupings simplify the task of determining how the jobs relate to each other (see Figure 2-2).

The groups can be established in this example through simple logic. In a complicated project, the diagrammer must be familiar with the details of the project or obtain assistance from experts. The experts would not only define the major categories, but would establish the relations between jobs. This procedure is the same as that used in developing the work breakdown structure (WBS). See Chapter 7 for further discussion of the WBS.

4. Determine relations between jobs.
The basic relations between jobs proceed from the planning assumptions made by the project planner. Note

MAJOR SECTIONS OF WORK

Stock
Packaging
Sales
Distributors
Advertising

Stock

Order stock
Package stock
Ship stock to distributors

Packaging

Design package
Set up packaging facility

Sales

Organize sales office
Hire sales personnel
Train sales personnel

Distributors

Select distributors
Sell to distributors

Advertising

Select advertising agency
Plan advertising campaign
Conduct advertising campaign

Figure 2-2 Groups representing major sections of work.

that these are preliminary plans at this point. Additional information will be obtained as the plan is refined. After the data are set up, someone knowledgeable would review the data.

Establishing job relations for a project generally involves a substantial amount of discussion regarding how activities occurring in different departments should be connected in the diagram. The resulting agreement on the flow of work should promote understanding of how various efforts and responsibilities tie together in the overall project.

5. Draw subdiagrams.   While the relations between jobs are being determined, it may be helpful to draw a subdiagram (Figure 2-3) for each group of jobs. Another way may be to determine the relations between jobs before starting on the subdiagrams. A combination of both schemes may also be useful for this step. When the subdiagrams have been developed, they are combined to form a complete diagram.

6. Draw complete diagram(s).   The last step in diagramming the project is to draw the complete network. This may involve:

- Drawing a preliminary diagram.
- Making necessary revisions to assure that the job relations are shown properly with no overrestrictions and looping, branching, and so on.
- Drawing a final diagram that provides a clear graphic display (Figure 2-4).
- Checking the final diagram for accuracy and effectiveness.

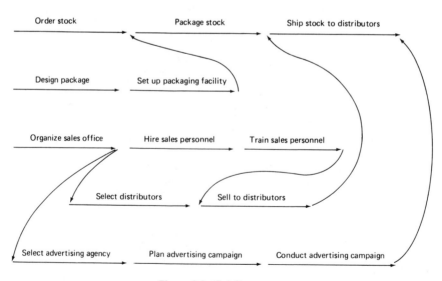

Figure 2-3 Subdiagram.

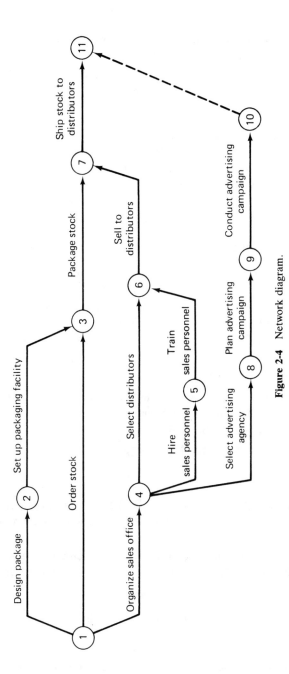

Figure 2-4 Network diagram.

It is possible that several preliminary diagrams will be drawn before the final diagram is completed. One preliminary diagram would ordinarily suffice, but projects containing input of several activities may represent different stages of network development. However, a second diagram emphasizes the idea that two people diagramming the same project will formulate different patterns and can produce equivalent results from the same jobs and job relations. Different planners might also establish different plans, and that will be reflected in their diagrams.

Arrow diagramming is a critical part of the planning phase of a project management cycle. Although the diagram will be only as good as the plans of those who developed it, it is an excellent technique. It may also be considered as a device for improving the plans by laying out the plan in clear and unambiguous detail. In this manner, planning assumptions and decisions are subjected to the test of logic.

Planning diagram check list.   List of job descriptions:

- Is the job detail suitable for the user (top management, other management, professional, and technical personnel)?
- Are the jobs stated as basic functions that can be understood by the intended audience?
- Does the list include all the work required for the project?

Preliminary diagram(s)

- Is the relation of each job specified accurately? What precedes it? What follows it? What can be done concurrently?
- Are there overrestrictions? looping? branching?
- Is there one starting point? one ending point?

Final diagram

- Is the project graphically displayed in a clear, easy-to-understand form? Is the flow of work simple to follow? Are groups of closely related jobs easy to identify?
- Is the network properly identified? Does each pair of numbers apply to only one arrow? Is the numbering scheme consistent with the computer program to be used?

The final questions relate to the complete planning diagram. Will the network serve:

- As an adequate base for the initial scheduling of the project?
- As an effective communication medium for evaluating, establishing, and controlling the project?

SUMMARY

Planning is the most important step in the project management process. Translating the jobs (and their sequence) into a graphical diagram (or a network) needs the most time. It is very common to hear someone respond, "I always do this in my head," upon being approached with the network analysis technique for planning; however, complexities of current projects dispel that approach. A disciplined analysis for planning a project is essential. Modern business methods can no longer rely on hunch and intuition as the basis of decision making. However, this technique does not replace good judgment but, rather, relies on good judgment to be an effective management tool.

Planning starts with having a complete understanding of the objectives. Then the following steps need to be taken:

1. Predetermine all of the activities or jobs that must be done to complete a project.
2. Develop a work breakdown structure.
3. Determine the sequence in which the jobs are to be done.
4. Develop the planning diagram.

If all of the steps above are performed in a reasonable manner, the diagram will become a valuable planning tool for the personnel involved in preparing the next phases of the project management cycle and for implementation of the project.

A graphic analysis provides a picture of the scope of the project. With this picture, there exists a vehicle for evaluating alternative strategies and objectives, a means of defining the interrelationships among jobs, and an excellent device for instructing personnel in the details of a project.

3
CPM/Project Scheduling

As stated previously, there is a planning phase and a scheduling phase in network planning. The planning phase, which was covered in Chapter 2, involves the use of the arrow diagram for analytical purposes. The purpose of this section is to describe how a project can be scheduled once the planning cycle is completed.

The first step in scheduling is to obtain time estimates for each job in the project. It is vital that the time estimates obtained be realistic in order to produce a good schedule for meeting deadlines and avoiding unnecessary project costs.

After the time estimates are obtained, the timing calculations begin. Optional starting and finishing times for the work items are developed as well as determining the critical work activities. This information is the forerunner of a complete schedule, making possible more effective use of personnel, eliminating unnecessary overtime, scheduling scarce equipment for maximum utilization, and possibly most important, controlling the cost of the project within its planned budget.

TIME ESTIMATES

A time estimate is obtained for each job in the project immediately after completing the initial arrow diagram. It may be accomplished before completing the planning phase, because at that point adjustments can be made to the diagram, if necessary, to meet management objectives. The time estimates may also be modified after the project begins. Usually, this becomes necessary if there is a delay in the work that will extend the duration of the project or if work is progressing more rapidly than was indicated by the original estimates.

The network diagram normally uses a single time for each job. The time estimate, usually determined by an experienced person, is the amount of time that the job will require under a specified set of conditions. The conditions may be as follows:

- *Normal:* The usual amount of labor, equipment, and so on, will be used.
- *Expedited:* Might involve the use of one or more factors, such as overtime, extra personnel, and additional equipment. The result is added cost.

First Estimates (Planning Stage)

The first estimates are made on the basis of how much time the job should actually take if everything proceeds on a normal basis. Persons familiar with the work to be performed should make the time estimates based on their best judgment. As network planning involves as accurate detail as possible, it is necessary to avoid any tendency toward padding the estimate to provide a cushion for contingencies, or underestimating because of undue optimism.

When the initial time estimates have been established, they are added to the arrow diagram (see Figure 3-1). There are certain rules to follow in applying the estimates on the diagram:

- Place time estimates on the bottom side of the arrow.

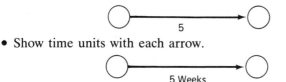

- Show time units with each arrow.

- Use whole numbers.
  - **a.** Manual calculations to develop schedules require whole numbers.
  - **b.** Most computer programs using the critical path technique for calculating schedules require whole numbers. (However, there are diagramming techniques, such as PERT, in which decimals are accepted.)

Second Estimates (Planning Stage—If Project Duration Must Be Reduced)

When the estimated duration of the project has been determined, it is examined relative to prior objectives. If the project duration, assuming "normal" conditions, is not satisfactory, new estimates must be made for some jobs.

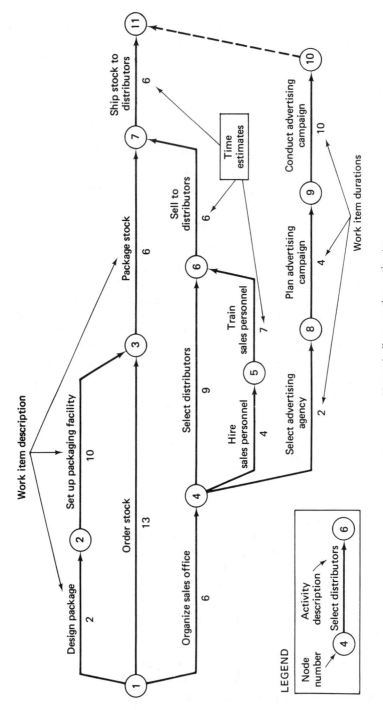

Figure 3-1 Network diagram: time estimates.

The method for determining the initial duration and which jobs must be shortened is related to the critical path technique for calculating the schedule and will be discussed later.

Third Estimates (Changes Occurring on Critical Path)

As work progresses on the project, some of the jobs may be completed before or after the scheduled time. If these changes are on the critical path and affect the project duration, modifications in the time estimates and the schedule will be required. When critical path jobs are finished ahead of schedule, it may be desirable to reschedule succeeding jobs so that an advantage can be gained by reducing the project duration. If jobs on the critical path are not completed on schedule, adjustments must be made in time estimates and the schedule to assure that the project will be completed within the scheduled project deadline.

Persons familiar with the project who are supplying the estimates may be influenced by previous experiences, creating a bias in their estimates. To offset this bias, a method using three time estimates can be adopted. The three time estimates—optimistic, most likely, and pessimistic—for each activity are used to offset the bias that may be present in one time estimate. The range of the time estimates also gives some indication of the scheduling risk involved; for example, a wide spread between the estimates shows considerable uncertainty about the time actually required to accomplish an activity.

Most likely time (or normal time).   The first time estimate requested is the most likely time. It is the time that would be most frequently required if the activity were repeated many times under similar conditions. It is also the normal job time estimate that would be shown on the arrow diagram.

Optimistic time.   The shortest possible time required for completing an activity is the optimistic time. Here it is assumed that everything goes as planned: deliveries of material occur on schedule, machines operate without major breakdowns, personnel perform work within work standards, and the like.

Pessimistic time.   The maximum possible time required to complete an activity is termed the pessimistic time. This is the time required for doing something if just about everything goes wrong: in short, the worst possible situation, including delays, accidents, equipment delivery difficulties, bad weather, and so on.

From these three pieces of information, an expected time is derived for the project with this formula:

$$\text{Expected time} = \frac{(\text{Optimistic time}) + 4(\text{Normal time}) + (\text{Pessimistic time})}{6}$$

This formula represents a weighted average of these three estimated times with two-thirds of the weight given to the normal time, one-sixth to the pessimistic, and one-sixth to the optimistic.

Those experienced in using the three estimates have found that the expected time will be biased toward the pessimistic time. As the one-estimate approach usually has a contingency built in, it will usually provide about the same estimated value as that provided by the three-estimate approach. One provides a good check for the other.

MANUAL TIMING CALCULATIONS

After time estimates are obtained, the timing calculations can be made. For a small project, these calculations are usually performed manually. We will use a sample problem (the planning diagram shown in Figure 3-1) to illustrate how to perform manual calculations. The time estimates are shown on the diagram. The problem consists of determining the earliest start times, latest finish times, and total float for a product introduction project.

Earliest Start Time

The term *earliest start time* at a node means the earliest time that any job can be started from that node. Jobs that begin at Node 1 are assumed to start at Time 0. For a given job, the earliest start time can be determined by adding the time estimate for the preceding job to the earliest start time for the preceding job (see page 36). The time elapsed between Time 0 and the earliest start time at a node thus represents the shortest period of time that will permit the completion of all preceding jobs that lead into that node.

Guides

The following guidelines should be considered when calculating the earliest start times for a project:

- The calculation of earliest start times begins with Node 1, the beginning of the arrow diagram (Time 0), and continues through each node to the end of the diagram.
- If only one arrow leads into a node, the earliest start time for jobs starting at the node is determined by adding the earliest start time for the preceding job to the time estimate for the preceding job (note Node 8):

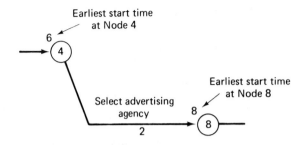

- If more than one arrow leads into a node, the earliest start time calculation is made through each of the arrows. The largest total is the earliest start time for the node, as noted at Node 7:

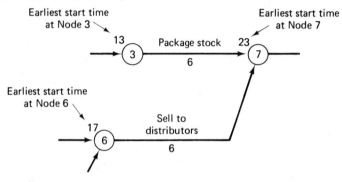

Application

Several representative calculations of earliest start times for the product introduction project are as follows:

		Earliest Start Time
Node 1	The earliest start time at Node 1 is set as time 0.	0
Node 2	Design Package requires 2 weeks. The earliest start time for Node 2 is, therefore:	
	0 Earliest start time at Node 1 +2 Time estimate for Job 1,2 2 Earliest start time at Node 2	2
Node 3	Two arrows lead into this node:	
	Set Up Packaging Facility requires 10 weeks: $2 + 10 = 12$	
	Order Stock requires 13 weeks: $0 + 13 = 13$	
	The larger total is used.	13

		Earliest Start Time
Node 4	Organize Sales Office requires 6 weeks.	
0	Earliest start time at Node 1	
+6	Time estimate for Job 1,4	
6	Earliest start time at Node 4	6

By using the same approach to calculate earliest start times for the remaining nodes, the earliest start times for all nodes are calculated and are noted in Figure 3-2.

Project Duration

At this point it can be seen that the jobs on the longest path in the diagram (the critical path) total 29 weeks for completion. This is the minimum project duration for this particular plan with these time estimates.

Latest Finish Time

The term *latest finish time* is the latest time that a job leading to a node can be completed without lengthening the duration of the project. The latest finish time is required in identifying the activities on the critical path and in calculating float for jobs in the network (see pages 40 and 41).

Guides

The following guidelines should be applied in determining the latest finish times for a project:

- The project duration must first be determined by calculating the early start times.
- The calculation of latest finish times involves working from the end node back through each node to the first node in the project. The project duration is the latest finish time for the last job.
- If the duration for a project is 29 weeks and Job 7,11, which requires 6 weeks, is the last job in the project, the latest finish time for jobs coming into Node 7 is 23.

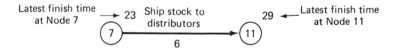

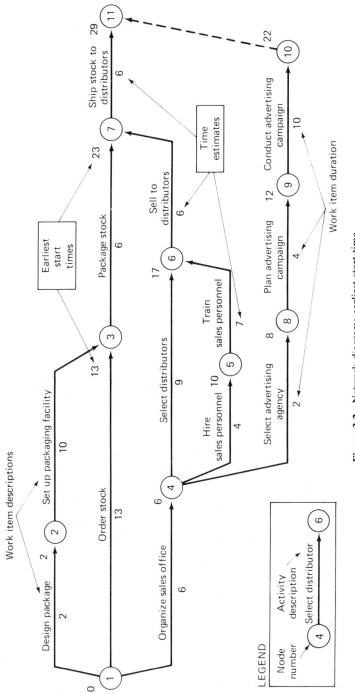

Figure 3-2 Network diagram: earliest start time.

The latest finish time for Node 7 is determined by subtracting the Job 7,11 time estimate from the project duration (the latest finish time at Node 11).

$$
\begin{array}{ll}
29 & \text{Latest finish time at Node 11} \\
-6 & \text{Time estimate for Job 7,11} \\
\hline
23 & \text{Latest finish time at Node 7}
\end{array}
$$

If more than one arrow originates at a node, the calculation of latest finish time is made via each arrow and the *smallest* result is used.

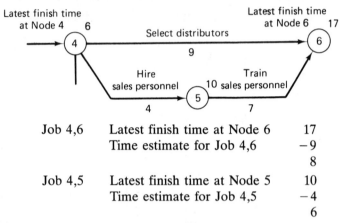

$$
\begin{array}{lll}
\text{Job 4,6} & \text{Latest finish time at Node 6} & 17 \\
 & \text{Time estimate for Job 4,6} & -9 \\
\hline
 & & 8 \\
\text{Job 4,5} & \text{Latest finish time at Node 5} & 10 \\
 & \text{Time estimate for Job 4,5} & -4 \\
\hline
 & & 6
\end{array}
$$

The latest finish time at Node 4 is 6 days, the smaller of the two results. Were the latest finish time for jobs coming into Node 4 set at 8 days, there would not be enough time remaining to complete Job 4,5 by its required latest finish time of 10.

Application

Several representative calculations of the latest finish times for the new product introduction project are as follows:

		Latest Finish Time
Node 11	The latest finish time for the end node is the project duration.	29
Node 10	Job 10,11 (Dummy) is 0-day duration.	
29	Latest finish time for Node 11	
− 0	Time estimate for Job 10,11 (Dummy)	
29	Latest finish time for Node 10	29
Node 9	Job 9,10 requires 10 days	
	29 − 10 = 19	19

Calculate the remaining latest finish times in the same manner. The latest finish times for all nodes in the project are noted in Figure 3-3.

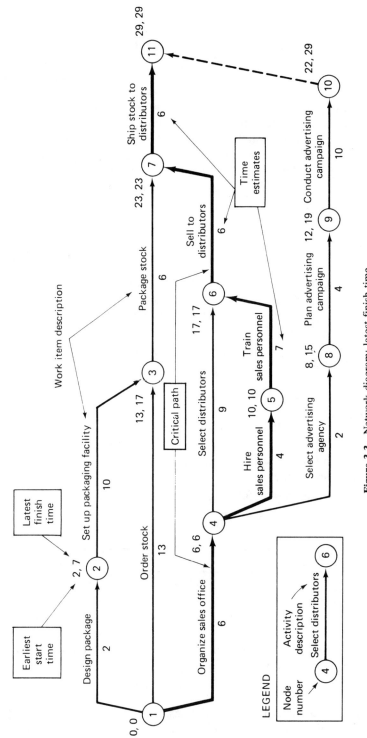

Figure 3-3 Network diagram: latest finish time.

When the earliest start times and latest finish times of all the activities have been determined, the critical path, the longest path on the network in terms of time requirements, can be identified by calculating the "total float" of each job. The method for calculating total float is shown in the next section.

The emphasis on the critical path may, at first, result in a tendency to pay too little attention to the jobs on the *subcritical paths*. These jobs are also important to the project, and their progress along each path must be monitored with the same degree of attention as the critical items during the progress of the project. Delays, or slippage, on a subcritical item may cause it to be the critical path, and if not controlled, it may lengthen the estimated duration of the project.

TOTAL FLOAT

From the scheduling viewpoint, one of the major advantages of network analysis is that it identifies the jobs that have optional starting and finishing dates. These jobs have *total float*, which is the difference between the time available for performing a job and the time required for doing it.

Computation

Total float is equal to:

- Time *available* for performing the job − Time *required* for performing the job.
- Time *available* for performing the job = Latest finish time (LF) − Earliest start time (ES).
- Time *required* for performing the job = Time estimate.

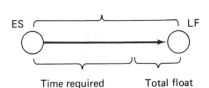

For example, using this formula, the total float for Job 4,6 on the product introduction project would be

$$(17 - 9) - 6 = 2 \text{ days of total float}$$

Another way to express total float can be seen in the illustration below. The time available for Job 4,6 is the difference between the latest finish time

and the earliest start time. The time estimate for Job 4,6 is subtracted from the
time available to determine the total float.

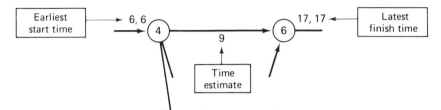

Total float figures are vital in scheduling a project. They indicate which
jobs have no float and cannot be delayed without extending the length of the
project. Equally important, they show which jobs have some leeway in scheduling (see Figure 3-4). You will note that jobs which have no float are marked
with an asterisk (*). These jobs are on the critical path, and their sum makes up
the longest path on the network in terms of time requirements.

It must be pointed out, however, that the total float figures must be used
with care. For example, there are 4 days of total float available for Job 1,3 of the
product introduction project. If this total float is used by scheduling Job 1,3 to
begin at time 4, the 4 days of float will not be available for Job 3,7, which is also
shown to have 4 days of total float. Job 3,7 must then be scheduled to begin at

Activity (node numbers)	Description	Latest finish time	(−)	Earliest start time	(−)	Time estimate for completing job	=	Total float
1, 2	Design package	7	(−)	0	(−)	2	=	5
1, 3	Order stock	17		0		13		4
1, 4	Organize sales office	6		0		6		0*
2, 3	Set up packaging facility	17		2		10		5
3, 7	Package stock	23		13		6		4
4, 6	Select distribution	17		6		9		2
4, 5	Hire sales personnel	10		6		4		0*
5, 6	Train sales personnel	17		10		7		0*
6, 7	Sell to distributors	23		17		6		0*
4, 8	Select advertising agency	15		6		2		7
8, 9	Plan advertising campaign	19		8		4		7
9, 10	Conduct advertising campaign	29		12		10		7
7, 11	Ship stock to distributors	29		23		6		0*
10, 11**	Dummy	29		22		0		7

* Critical path.
** Dummy—has no time duration.

Figure 3-4 Computation of total float.

time 17. It cannot be scheduled to begin at time 13, its earliest start time, because the total float has already been used on Job 1,3.

Total float figures thus indicate all the places where float is available and may indicate the same float at more than one place. There are two other types, free float and independent float, which tend to clarify this situation. These types of float will be defined later.

Where to Use Float

At what point should total float be taken when jobs are being scheduled? This decision is based on such factors as requirements for various resources, effect on cost, stability of employment, and cost penalties that may be incurred if certain jobs are delayed. For example, in the project introduction project, the float available on Job 9,10 (Conduct Advertising Campaign) would probably be used to delay its completion until the last possible moment. The advertising campaign should coincide more with the sales made by the distributors. Scheduling decisions of this kind within the framework of the float considerations can be made throughout the project.

FREE AND INDEPENDENT FLOAT

Float is defined as the time the performance of a job may be delayed without delaying overall project completion. If a job is on the critical path, it has no float. On all other paths in the diagram, some slippage is permissible without delaying project completion. Float figures make it possible to identify jobs that have optional starting and finishing dates. The determination of float for these jobs is important in order to identify the various scheduling operations that the scheduler may use to better allocate money, labor, and other resources to the project.

There are three types or definitions of float: total, free, and independent.

Total float is the least restrictive definition of float. The calculation of total float tells the scheduler all the jobs in the network on which float might be taken. It must be remembered that the same float will be shown on the chain of activities along the same path.

Free-float figures have the advantage that each occurrence of float is recorded only once in a project. The effect of free-float computations is to "push" the float to the last job in the chain of activities in which the float occurs.

Independent-float data isolate the float that is available on one job and one job only. Independent float is calculated on the assumption that all previous jobs are scheduled as late as possible and all succeeding jobs as early as possible.

Each type of float has some drawbacks in application, but when all three are considered together, they provide valuable data for effectively scheduling a project. Before defining these in more detail, let us define the major factors in the float calculations:

1. *Earliest start time* (ES): earliest time any job can be scheduled, assuming that all previous jobs were completed on time. The project duration is equal to the earliest start time of the ending node of the last activity in the project.

2. *Latest finish time* (LF): latest time that a job can be completed without lengthening the duration of the project. The latest finish time is required in identifying the critical path and in calculating float for jobs in the network.

3. *Total float* (TF): the difference between the maximum time available for performing a job and the estimated time required for doing it.
   - The same total float may be recorded for more than one activity in a path.
   - If all total float is taken, all following activities in that path become critical.

4. *Free float* (FF): the delay possible in an activity if all preceding jobs start as early as possible while allowing all subsequent jobs to start at their earliest time.

Formula for Free Float for Job X

$$\text{Free float} = \begin{bmatrix} \text{Earliest start time} \\ \text{for jobs immediately} \\ \text{following Job } X \end{bmatrix} - \begin{bmatrix} \text{Earliest time} \\ \text{time for Job } X \end{bmatrix} - \begin{bmatrix} \text{Duration} \\ \text{of Job } X \end{bmatrix}$$

Free float is recorded selectively. The effect of free-float calculations is to push the float to the last activity in the chain in which the float occurs.
   - Free float is recorded for one activity only; thus, free float provides a safety factor.
   - Available float is shown only on the last activity.

5. *Independent float* (IF): the delay possible in an activity if all preceding jobs are completed as late as possible and all subsequent jobs are to be started at the earliest possible start time. The purpose of calculating independent float is to isolate where float must be used on one job and is not available to any other job.

Formula for Independent Float for Job X

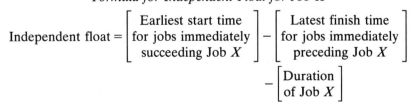

$$\text{Independent float} = \begin{bmatrix} \text{Earliest start time} \\ \text{for jobs immediately} \\ \text{succeeding Job } X \end{bmatrix} - \begin{bmatrix} \text{Latest finish time} \\ \text{for jobs immediately} \\ \text{preceding Job } X \end{bmatrix} \\ - \begin{bmatrix} \text{Duration} \\ \text{of Job } X \end{bmatrix}$$

- If the result is negative, the independent float for Job $X = 0$.
- Independent float isolates the float that is available on one job and one job only. The calculations are based on all previous jobs and are scheduled as late as possible, and all succeeding jobs are scheduled as early as possible.

Examples of free float and independent float using the product introduction project are shown in Figure 3-5.

Calculations for Latest Start and Earliest Finish

Determining optional starting and finishing times for project activities can be of some benefit to the scheduler. Both optional starting and finishing times are related to the total float value of each project activity.

Latest start = Earliest start + Total float
Earliest finish = Latest finish − Total float

The schedule tabulation for the product introduction project shown in Figure 3-6 illustrates all the optional dates: the earliest and latest starts as well as the earliest and latest finishes of the project jobs.

FREE FLOAT

i, j	Description of job X	Earliest start time of succeeding activities	(−)	Earliest start time of job X	(−)	Duration of job X	=	Free float
4, 8	Select advertising agency	(Job 8, 9) plan advertising campaign 8		(Job 4, 8) select advertising agency 6		(Job 4, 8) 2	= =	0
8, 9	Plan advertising campaign	(Job 9, 10) conduct advertising campaign 12		(Job 8, 9) plan advertising campaign 8		(Job 8, 9) 4	= =	0
9, 10	Conduct advertising campaign	(Job 10, 11) dummy 22		(Job 9, 10) conduct advertising campaign 12		(Job 9, 10) 10	= = 0 =	7*
10, 11	Dummy	(End of job) 29		(Job 10, 11) dummy 22		(Job 10, 11) 0	=	7

*Where a dummy is at the end of the chain of activities, the free-float value is transferred to the preceding job activity.

INDEPENDENT FLOAT

i, j	Description of job X	Earliest start time of succeeding activities	(−)	Latest finish time of preceding activity	(−)	Duration of job X	=	Independent float
1, 4	Organize sales office	(Job 4, 6) select distributors (Job 4, 5) hire sales personnel (Job 4, 8) select advertising agency 6		Start of job 0		(Job 1, 4) 6	=	0
2, 3	Set up packaging facility	(Job 3, 7) package stock 13		(Job 1, 2) design package 7		(Job 2, 3) 10	= −4 =	0*
4, 6	Select distributors	(Job 6, 7) sell to distributors 17		(Job 1, 4) organize sales office 6		(Job 4, 6) 9	=	2
5, 6	Train sales personnel	(Job 6, 7) sell to distributors 17		(Job 4, 5) hire sales personnel 10		(Job 5, 6) 7	=	0

*A negative independent float is equal to zero independent float.

Figure 3-5   Calculating free float and independent float.

i, j	Description	Time (weeks)	Earliest Start	Earliest Finish	Latest Start	Latest Finish	Float Total	Free	Independent
1, 2	Design package	2	0	2	5	7	5	0	0
1, 3	Order stock	13	0	13	4	17	4	0	0
1, 4	Organize sales office	6	0	6	0	6	0	0	0
2, 3	Set up packaging facility	10	2	12	7	17	5	1	0
3, 7	Package stock	6	13	19	17	23	4	4	0
4, 5	Hire sales personnel	4	6	10	6	10	0	0	0
4, 6	Select distributors	9	6	15	8	17	2	2	2
4, 8	Select advertising agency	2	6	8	13	15	7	0	0
5, 6	Train sales personnel	7	10	17	10	17	0	0	0
6, 7	Sell to distributors	6	17	23	17	23	0	0	0
7, 11	Ship stock to distributors	6	23	29	23	29	0	0	0
8, 9	Plan advertising campaign	4	8	12	15	19	7	0	0
9, 10	Conduct advertising campaign	10	12	22	19	29	7	7	0
10, 11	Dummy	0	22	22	29	29	7	7	0

Figure 3-6 Schedule tabulation.

An expedient in translating tabulated schedule dates into calendar dates is the date calculator, one version of which is shown in Figure 3-7. Date calculators can be procured through bookstores and stationery stores. Assuming a designated starting date of April 2, 1990 (Time 0), we can illustrate its use as follows (refer to Figure 3-6):

Item Description	Earliest Start	Latest Finish
Organize sales office	4/2/90	5/14/90
Select advertising agency	5/14/90	7/16/90
Sell to distributors	7/30/90	9/10/90

The computerized version of the manually calculated tabulation is shown in Figure 3-8. The basis of this computerized schedule is the same as that used

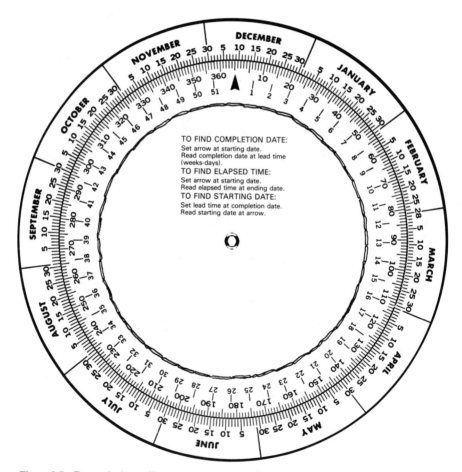

TO FIND COMPLETION DATE:
Set arrow at starting date.
Read completion date at lead time (weeks-days).
TO FIND ELAPSED TIME:
Set arrow at starting date.
Read elapsed time at ending date.
TO FIND STARTING DATE:
Set lead time at completion date.
Read starting date at arrow.

Figure 3-7 Date calculator. (Designed and produced by PERRYGRAF, Northridge, CA 91324.)

for the manual schedule tabulation: unique numbering system (node numbers), description of each activity, and duration of each activity. Comparing Figure 3-8 with Figure 3-7, you will note discrepancies in the latest finish dates. Computerized schedule tabulation shows Friday as the latest finish date, whereas the schedule using the date calculator will show a Monday date.

CONSTRUCTING A BAR CHART SCHEDULE

The network diagram is excellent for laying out and planning a job. It is very useful in visualizing the status of the job by highlighting the critical items and potential bottlenecks. Its importance lies in showing the interrelationships among the project. However, for scheduling jobs, the network analysis technique does have some limitations:

1. Often, the network is difficult to interpret.
2. A great deal of time is usually needed to prepare changes. These changes will also require a great deal of time for making modifications to the network diagram.
3. A network makes it difficult to note estimated costs versus actual costs.
4. Individual skills are not recognized.

A bar chart recognizes the scheduling limitations when using network analysis; however, its construction needs the information derived from the network analysis calculations. To illustrate this, the bar chart time schedule for

REPORT TYPE   : SCHEDULE TABULATION : Normal
TIME NOW DATE: 1/APR/90   PRODUCED BY: P1211

Description	Early Start	Late Start	Early Finish	Late Finish	Duration	Float
DESIGN PACKAGE	2/APR/90	7/MAY/90	13/APR/90	18/MAY/90	10	25
SET UP PACKAGING FACILITY	16/APR/90	21/MAY/90	22/JUN/90	27/JUL/90	50	25
ORDER STOCK	2/APR/90	30/APR/90	29/JUN/90	27/JUL/90	65	20
PACKAGE STOCK	2/JUL/90	30/JUL/90	10/AUG/90	7/SEP/90	30	20
SHIP STOCK TO DISTRIBUTORS	10/SEP/90	10/SEP/90	19/OCT/90	19/OCT/90	30	0
ORGANIZE SALES OFFICE	2/APR/90	2/APR/90	11/MAY/90	11/MAY/90	30	0
SELECT ADVERTISING AGENCY	14/MAY/90	2/JUL/90	25/MAY/90	13/JUL/90	10	35
HIRE SALES PERSONNEL	14/MAY/90	14/MAY/90	8/JUN/90	8/JUN/90	20	0
PLAN ADVERTISING CAMPAIGN	28/MAY/90	16/JUL/90	22/JUN/90	10/AUG/90	20	35
TRAIN SALES PERSONNEL	11/JUN/90	11/JUN/90	27/JUL/90	27/JUL/90	35	0
CONDUCT ADVERTISING CAMPAIGN	25/JUN/90	13/AUG/90	31/AUG/90	19/OCT/90	50	35
SELECT DISTRIBUTORS	14/MAY/90	28/MAY/90	13/JUL/90	27/JUL/90	45	10
SELL TO DISTRIBUTORS	30/JUL/90	30/JUL/90	7/SEP/90	7/SEP/90	30	0

Figure 3-8 Computerized schedule tabulation.

the sample problem, "A New Product Introduction," is shown in Figure 3-9. The procedure for constructing this bar chart is as follows:

1. Use the earliest start time of each project activity.
2. The length of each bar is the duration of each activity.
3. Plot one activity per line. In some cases it may be advantageous to plot a number of activities on one line. If the critical path items are shown on one line, a change to the planned schedule of each of the items and the effect on the project duration can be noted immediately.

SUMMARY

An important part of the project scheduling phase is time estimating, which involves getting a time estimate for each job in the project. The time estimate represents the amount of time an experienced person thinks the job will require under specified conditions. The first set of time estimates is generally made on

SUMMARY

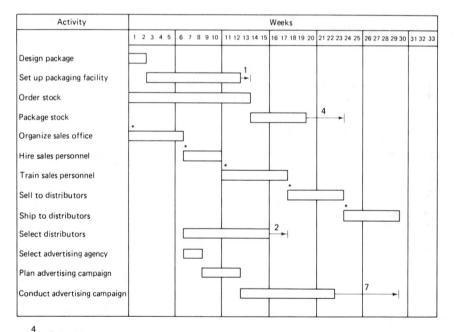

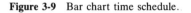

Figure 3-9   Bar chart time schedule.

the assumption that the project will be accomplished on a normal basis (employing readily available resources and using a minimum of overtime and other special measures). If the project duration based on the first set of time estimates does not meet project objectives, a second set of estimates is obtained. Reductions are made in time estimates by expediting jobs on the critical path; in this way, the project duration is brought within the project deadline.

During the project, revised time estimates may be necessary if the progress of work varies from the schedule. When delays occur, time must be removed from some remaining jobs to assure that the project will be completed on schedule. If work is progressing ahead of schedule, adjustments may be made to shorten the overall duration of the project.

The timing calculations provide information necessary to schedule a project effectively. For small projects, calculations can be made manually; for larger projects, a computer is used. In the first series of calculations, the earliest start time at each node in the network is determined. The earliest start time at a node is the earliest time any job can be scheduled to start from that node.

The computation of the earliest start times also yields the project duration. This duration is checked against project objectives. If it does not meet project objectives, the duration must be compressed by expediting some of the critical jobs on the project.

The latest finish time for each node on the network is then calculated. This is the latest time that jobs can be completed without lengthening the duration of the project. The earliest start and latest finish times permit another calculation that yields the total float for each job—the difference between the time available for performing a job and the time required for performing it. The total float calculations make it possible to identify the critical path readily.

The benefits to be derived from the critical path timing calculations are principally (1) the establishment of the project duration for the plan, (2) the identification of the longest path (critical path) through the project, and (3) the identification of jobs for which there is scheduling flexibility without lengthening project duration.

Float figures make it possible to identify jobs that have optional starting and finishing dates. Total float is the least restrictive definition of float. The calculation of total float tells the scheduler all the jobs in the network on which float might be taken. It must be remembered that the same float may be indicated on more than one job.

Free-float figures have the advantage that each occurrence of float is recorded only once in a project. The effect of free-float computations is to push the float to the last job in the chain of activities in which the float occurs.

Independent-float data isolate the float that is available on one job and one job only. Independent float is calculated on the assumption that all previous jobs are scheduled as late as possible and all succeeding jobs as early as possible.

Each type of float—total, free, and independent—has drawbacks in application, but when all three are considered together, they provide valuable data for effectively scheduling a project.

With the planning diagram, the list of critical activities, and the remainder of the jobs with their available float, the job can be scheduled. The schedule tabulation is a continuation of the timing calculations using the float values. The computerized schedule is generated by utilizing the same data as those used for calculating the manual schedule. However, more detail can be developed for analysis purposes when using computer calculations.

4
Project Control

Project control, the third phase of the project management cycle, generally consists of continuously monitoring the progress of each project item, then, to keep the project on the planned schedule, taking the necessary action on those items shown to be "drifting." The specific steps to be taken for effective project control are as follows:

1. Monitor.
2. Assess.
3. Resolve.
4. Communicate.

Monitoring involves a periodic survey of the status of the work activities. This is done by entering the actual start and/or completion dates (and any revised duration) into the computerized schedule and generating a new computer schedule. The results of this technique are illustrated in Figure 4-1. This particular survey was taken 12 weeks into the project.

Figure 4-1 can also be used for the second step, assessing the status. (Note that Figure 4-1, updated schedule, shows the dates in the "Early Start" and "Early Finish" columns.) Priorities in the assessment process deal with those critical items that are drifting from the planned schedule. Further tasks concerning the contents of this illustration are covered in later portions of this chapter.

NEW PRODUCT INTRODUCTION

REPORT TYPE : PROJECT SCHEDULE (UPDATED) : Normal
TIME NOW DATE: 25/JUN/90   PRODUCED BY: P1211

Name	Description	Early Start	Late Start	Actual Start	Early Finish	Late Finish	Actual Finish	Duration	Float
034 -002	DESIGN PACKAGE	2/APR/90	2/APR/90	2/APR/90	13/APR/90	16/APR/90	13/APR/90	10	1
002 -003	SET UP PACKAGING FACILITY	16/APR/90	17/APR/90	17/APR/90	6/JUL/90	27/JUL/90	/	50	15
001 -003	ORDER STOCK	2/APR/90	23/APR/90	23/APR/90	20/JUL/90	27/JUL/90	/	65	5
003 -007	PACKAGE STOCK	23/JUL/90	30/JUL/90	/	31/AUG/90	7/SEP/90	/	30	5
007 -011	SHIP STOCK TO DISTRIBUTORS	24/SEP/90	10/SEP/90	/	2/NOV/90	19/OCT/90	/	30	-10
001 -004	ORGANIZE SALES OFFICE	2/APR/90	2/APR/90	2/APR/90	11/MAY/90	11/MAY/90	11/MAY/90	30	0
004 -005	HIRE SALES PERSONNEL	14/MAY/90	28/MAY/90	28/MAY/90	22/JUN/90	8/JUN/90	22/JUN/90	20	-10
005 -006	TRAIN SALES PERSONNEL	25/JUN/90	11/JUN/90	/	10/AUG/90	27/JUL/90	/	35	-10
004 -006	SELECT DISTRIBUTORS	14/MAY/90	14/MAY/90	14/MAY/90	13/JUL/90	27/JUL/90	/	45	10
006 -007	SELL TO DISTRIBUTORS	13/AUG/90	30/JUL/90	/	21/SEP/90	7/SEP/90	/	30	-10
004 -008	SELECT ADVERTISING AGENCY	14/MAY/90	14/MAY/90	14/MAY/90	25/MAY/90	25/MAY/90	25/MAY/90	10	0
008 -009	PLAN ADVERTISING CAMPAIGN	28/MAY/90	28/MAY/90	28/MAY/90	24/JUN/90	10/AUG/90	24/JUN/90	20	34
009 -010	CONDUCT ADVERTISING CAMPAIGN	25/JUN/90	13/AUG/90	/	31/AUG/90	19/OCT/90	/	50	35

Figure 4-1 Updated schedule (week 12).

Utilizing the planning and scheduling techniques described in this book makes it possible to resolve problems (such as delayed work items) on an early-warning basis. The computerized schedule projects finish dates and highlights those items that show behind-schedule situations with a negative ($-$) float value. The longer the negative value, the more critical the situation is.

SUMMARY BAR CHART

One of the best expedients used by those keeping track of job progress (and also used for management reports) is the bar chart. Actions taken on a daily basis are shown on the detailed bar chart schedule. For management review on a weekly or monthly basis, a summary bar chart is prepared, which is based on the detailed time schedule. A summary bar chart is part of the summary schedule, which is an excellent communication device to apprise management of the status of a project.

The bar chart introduces some features that may require explanation, either in the form of a legend noted on the chart or in a supplemental attachment. For example, the summary chart in Figure 4-2 requires the following explanation:

1. The rectangular bars (▭) represent the project phases (design, purchase order, fabrication, installation, tryout, etc.) of each major item in a timing

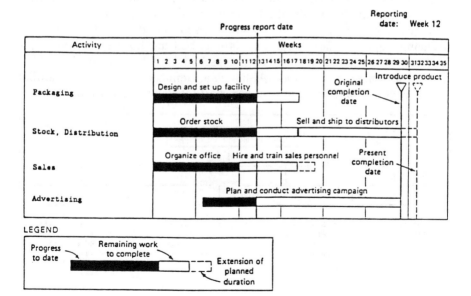

Figure 4-2 Summary bar chart.

sequence to meet the projected completion date. The portion that is completed is filled in (■).

2. Critical project event dates or milestones (▽) are left hollow until that event is completed and then are filled in (▼).

3. The planned program completion milestone (▽) is shown in the upper right of the chart. If the completion milestone requires a time change because of changes in scope, technical problems, late deliveries, and so on, the updated completion milestone will be noted with a dashed line (▽).

(The Activity column in Figure 4-2 is arranged into the work breakdown structure format as shown in Figure 4-3.)

The summary bar chart for management reports uses the late start schedule, in contrast to the progress schedule, which will use the early start schedule for implementing a project. Providing late start schedules to those responsible for implementation can be dangerous, as this schedule allows for no margin in the event of delays on critical items or even low float items, resulting in a potential project completion delay.

Using the late start schedule for reporting to management is useful, as only critical items are highlighted, thereby avoiding extraneous discussions on items that are not critical.

Figure 4-4 shows a computer-generated bar chart developed from the basic data. This report provides the same information as that provided on a manually prepared bar chart, but can be revised and expanded in various forms with less time and effort required than in manual preparation.

How the summary bar chart is summarized from the bar chart schedule, the manner in which the current status is described, and how off-schedule items are handled to maintain planned progress schedules will determine how effective it is as a management tool. Although a summary bar chart is somewhat similar in construction to a project bar chart schedule, there are basic differ-

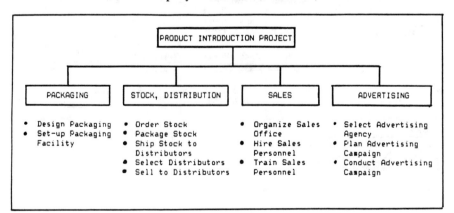

Figure 4-3   Work breakdown structure.

ences. For example, items in the status schedule are grouped according to the work breakdown structure approach. This approach should be shown and explained to management and project personnel, as it is an excellent expedient in project planning and control. In the product introduction project, the major groups (shown with the items under each group) would be as follows:

Packaging

- Design package.
- Set up packaging facility.

Stock, distribution

- Order stock.
- Package stock.
- Ship stock to distributors.
- Select distributors.
- Sell to distributors.

Sales

- Organize sales office.
- Hire sales personnel.
- Train sales personnel.

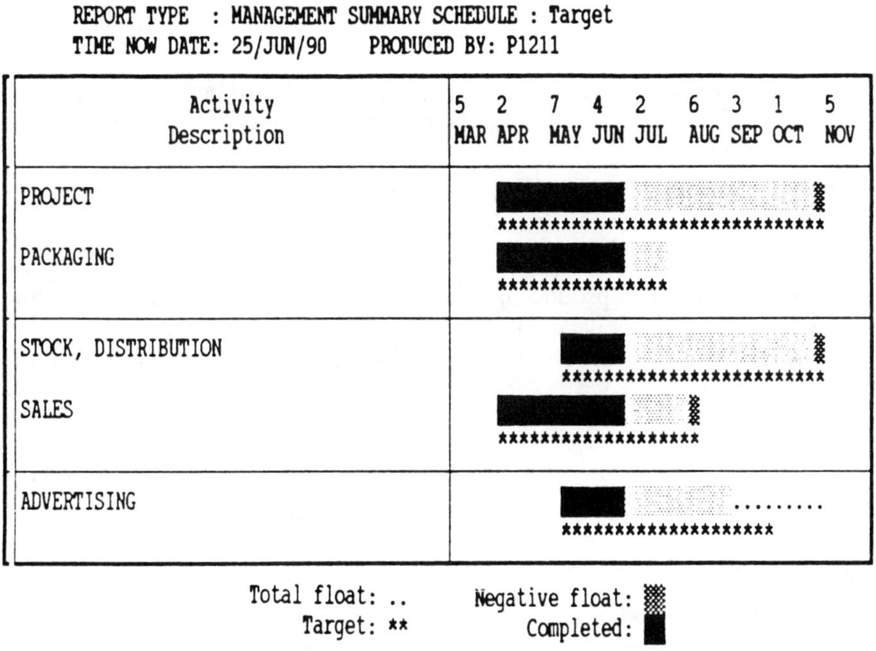

```
REPORT TYPE  : MANAGEMENT SUMMARY SCHEDULE : Target
TIME NOW DATE: 25/JUN/90    PRODUCED BY: P1211
```

Activity Description	5 MAR	2 APR	7 MAY	4 JUN	2 JUL	6 AUG	3 SEP	1 OCT	5 NOV
PROJECT									
PACKAGING									
STOCK, DISTRIBUTION									
SALES									
ADVERTISING									

```
      Total float: ..   Negative float: ▓
          Target: **        Completed: ■
```

Figure 4-4 Computerized summary bar chart.

Advertising

- Select advertising agency.
- Plan advertising campaign.
- Conduct advertising campaign.

PROJECT STATUS REPORT

Project control is predicated on an early start/early finish philosophy, the same as that initiated in project scheduling. This principle is used in the *project status report,* which is the discipline used to provide a periodic factual record as a basis for corrective action. The status report is used to assess, anticipate, evaluate, and then recommend to management the corrective action of any delay or drift.

The basic information (*monitor* step) needed for project control is derived from the project schedule, and the resulting status report will employ charts and graphs to portray the status of the project. Exception reporting (developed during the *assess* step) is one of the most important aspects of status reporting. For example, the following points need to be raised in the report:

1. Which jobs, specifically, are behind schedule? How much will this extend the project completion date if not corrected?
2. What are the reasons for the delays that have caused the project to fall behind schedule?
3. What steps have been or are being taken (or should be taken) to restore the situation to its initial place, and what results have been obtained or are expected?
4. What specific recommendations for further action were provided that will restore the corrected situation?

To illustrate the product introduction report summary bar charts, the following scenario was developed.

A review of the work activities at the end of the twelfth week indicated the following status:

	Based on Earliest Start Time	Total Float
Design Package and Set Up Packaging Facility	Behind 2 weeks	5 weeks
Order Stock	Behind 3 weeks	4 weeks
Hire and Train Sales Personnel	Behind 2 weeks	0 weeks
Plan and Conducts Advertising Campaign	Behind 4 weeks	7 weeks

Based on earliest start times and the total float available, the present status of the work activity group, *Hire and Train Sales Personnel,* affects the completion date at the present rate of progress. This project will be completed 2 weeks later than planned (unless corrective action is taken).

The steps in reporting to management through the summary chart are:

1. Show the actual program status compared to the plan.
2. Revise time estimates, if required. Portray graphically any extensions beyond the planned completion time.
3. Highlight critical items in the commentary below the bar chart.
4. Introduce options to achieve objectives. If there is evidence that the project completion will be delayed, show this possibility early in the project. If a delay in the project duration is inevitable, show the extent of delay on the bar chart.

The status report is a reflection of the program content in a summary outline form. It contains the status of key project items, an assessment of these items, and those items that should be given special attention. Specifically, a "full-blown" project status report will contain the following documents:

- Cover letter
- Executive highlights
- Summary of project
  a. Bar chart
  b. Project status
  c. Milestone report

The *cover letter* (Figure 4-5) is addressed to the management personnel whose activity is participating or who has other interests in the particular project. Its contents briefly explain the program of the job, anticipated completion dates of major events, brief statements concerning the status of critical items, and potential solutions to any problems.

Executive highlights (Figure 4-6) underline the status of the important aspects of the project. Highlights can be in the form of a "bullet" listing (•), which should consist of a brief sentence or two, without going into too much detail. In brief progress status reports, executive highlights may sometimes be incorporated in the cover letter or in the project summary.

The *project summary* (Figure 4-7) includes a bar chart of progress to date and a milestone report. The bar chart combines the work activities in a graphic form in such a manner that management can review the overall project and be spared the countless details associated with day-by-day activity. It applies the management by exception principle, which allows more attention to the critical items.

---

5 August 1991

Messrs. J. G. Beck
        B. L. Sarkis

Subject: 1 August 1991 Status Report
            Glass Coating Project

The Glass Coating Project is now anticipated to be completed on 1 February 1992 (one month later than planned schedule). A supply of promotional samples will be available at that time for sales and marketing to use at its product introduction reception.

Essential phases of the project—equipment facilities installation and procurement—are continuing on the original schedule; however, the two July trials conducted were inconclusive, which will require on-site equipment modifications and will delay tryout by about one month.

There is a high probability that these equipment changes will allow subsequent trials to be conducted with no delay.

For further information on the status, please review the project summary containing the bar chart, project status, and milestone report, which are attached.

Please advise if you have any comments or further questions.

                                        M. Pete Spinner

MS/lrd
Attachments

Figure 4-5 Cover letter.

The status commentary (or progress to date) below the bar chart explains in more detail the status of the project items shown in the bar chart. As the bar chart graphically portrays the status of the major activities, the status commentary is used to highlight the details of these activities, with particular attention to items that are critical.

The milestone portion of the status report includes the planned dates of project start and completion, deviations from these dates, and explanation of the variances. Important interim dates of the project are also shown, as these

---

GLASS COATING PROJECT
EXECUTIVE HIGHLIGHTS

- Fabrication of process equipment is over 75% completed, and about 60% of the assembly is completed. "Debugging" at Tech Center, prior to shipment to Tulsa, is expected to begin October 1.

- Equipment tryout will start November 15; by December 1, manufacturing expects to confirm ability to produce acceptable product.

- Based on successful tryouts in December and February, glass shipments are projected to start by February 1.

- Research and Development, Glass Tech Center, will provide Sales and Marketing with initial promotional samples for mailing campaign. The 3″ × 4″ seamed edge glass samples will be completed by December 1.

- Timetable for supplying other glass samples for the Marketing introduction program is based on outcome of tryouts in early December.

                                        Glass Engineering
                                        8-4

Figure 4-6 Executive highlights.

dates directly reflect the major milestone dates to be met for the timely completion of the project.

The project status report may be of one page or several pages. Although its format may be designed differently from the one shown here, the contents would be similar. However, there may be additional special features, depending on the project requirements. In any event, its contents will contain the essential data that management needs to appraise the project properly.

As the summary bar chart is a most important communication, we are providing further details on its contents, including the milestone concept.

Milestones

The milestone approach is a tool for project planning and project control and is especially effective with the assistance of a computer software program. Milestones are selected events that are of major importance toward achieving

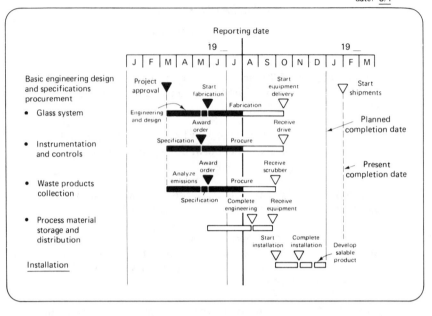

Progress to Date

There has been little development progress in the past two months. Most of the on-line trial malfunctions have been caused by mechanical difficulties, which are being corrected for the scheduled August trials. Procurement of facility equipment is on schedule, with no critical delivery problems existing at this time. Progress details of major items are as follows:

Glass process equipment — Fabrication, about 75% complete, is planned to be ready for "debugging" by October 15 at the Tech Center prior to shipment to Tulsa. Required for "debugging": the air control panel is now being fabricated and the electrical control panel order was awarded July 25. Both panels are expected to be completed and shipped to the Tech Center in September.

Exhaust and waste collection — Duct work should be completed by August 1 and sent to the Tech Center for assembly. Supplier of the special air scrubber confirms Tulsa delivery by October 15.

Process material storage, mixing, and distribution — Behind schedule condition due to lengthy negotiations. Supplier's quotation is considered excessive. To avoid production delay, Tulsa will assume responsibility and provide temporary systems.

> **Milestone Report**
> Schedules remain the same as last report, installation of equipment ex-
> pected to be completed November 15, as planned. Production start date of
> February 1 is contingent on successful tryouts in December and January.
>
> | | |
> |---|---|
> | November 15 | Complete equipment installation; start tryout. |
> | December 5 | Begin preparations for system tryouts. |
> | January 15 | Provide marketing with glass for trade shows, model homes, etc. |
> | February 1 | Start production shipments. |

Figure 4-7 Project summary: bar chart, project status, and milestone report for the glass coating project.

objectives. They are key events, usually showing the completion date of a major phase of the project, the delivery date of a major equipment item, or the date of a key management decision to make the project maintain its successful completion. These events may or may not be on the critical path. The milestone approach is used for reporting project status in summary form to higher management, as it essentially summarizes the status of the major events.

In Figure 4-8 the milestones for the product introduction project are displayed. These milestones highlight the important events of this project, leading to the project completion date. The milestones can also be shown on a network diagram (Figure 4-9), representing the key starting or completion dates of major events. The display of milestones on the network diagram can be helpful for portraying the planning strategy; however, in most instances the milestones can be identified only through the development of computer calculations (and then shown on the planning diagram).

As the computer-oriented milestone technique provides a condensed version or summary to help direct the project to a successful completion, it offers a "snapshot" of the project, as well as a number of advantages, such as:

- The milestone listing through the computer printout provides a precise form in which program progress can be monitored.
- The computer program can sort a milestone listing highlighting the items that need attention to ensure that the project can be completed on schedule.
- Additional milestones can be added during the course of the project. These are added to monitor a potential critical activity or to monitor the entire chain of preceding activities along that path to the point of the milestone event.

One technique used for monitoring the status of milestones is to "dam up" the time at the milestone, thereby not permitting any behind-schedule

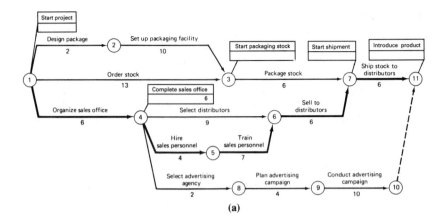

(a)

REPORT TYPE   : MILESTONE REPORT : Normal
TIME NOW DATE: 25/JUN/90   PRODUCED BY: P1211

Description	Milestone Date
START PROJECT	2/APR/90
COMPLETE SALES OFFICE	14/MAY/90
START PACKAGING STOCK	23/JUL/90
START SHIPMENT	24/SEP/90
INTRODUCE PRODUCT	5/NOV/90

(b)

Figure 4-8   (a) Planning diagram; (b) milestone report for the product introduction project.

activities to pass that point in time. In the case of behind-schedule situations, each group of behind-schedule activities can be isolated, examined, and recommendations made to get the project back on schedule. It is possible that as one group of activities is in the process of being resolved, another series requiring attention may present itself. Where the behind-schedule activities are quite prevalent, it is a constant "whittling" process. This approach is most effective when introduced in the early stages of the project. Corrective actions need to be taken at each milestone, as early as possible; otherwise, subsequent activities will be adversely affected and in all probability the objective will be missed.

The milestone approach is an important and effective tool and should be used by all project personnel in the pursuit of their projects.

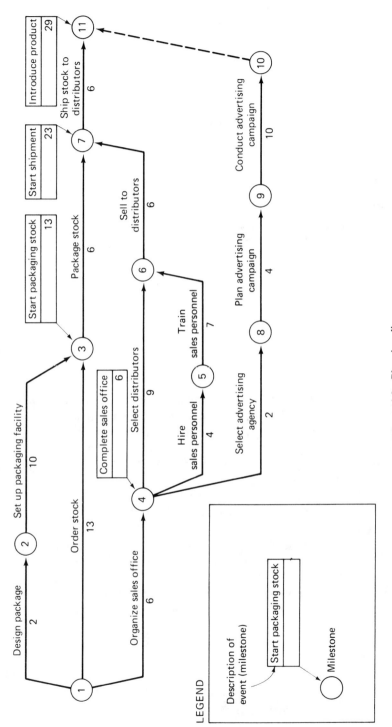

Figure 4-9 Planning diagram.

5
Project Costs

To many in management, the costs associated with a project may be more important than the timing aspects, and they have a great impact on how the project is carried out. Project costs are usually concerned with the flow of cash or movement of money throughout the various groups connected with the project. The owner or company officer financing the project is interested and must be made knowledgeable of these two main considerations:

1. What total amounts of money will be needed over the course of the project?
2. When will the money be needed to make payments for materials, labor, and other expenses over the duration of the project?

An accurate determination of the funds allocated to the project and control of these funds throughout the project are important to the financial integrity of an organization.

This chapter is devoted to three aspects of project costs:

- Project cost schedule
- Project cost control (indicated cost outcome)
- Cost minimizing (time/cost trade-offs)

Although these three aspects do not represent the total cost implications of a project, they do cover the more important cost areas. Persons directly associated with project operations should be aware that successful management of costs is essential to proper project management.

The initial effort involves the development of costs for each project item. Project costs are established for each work activity in the same manner as the timing for the project was developed. As is the case in developing timing estimates, it is best for those knowledgeable of the specific work items to furnish the cost estimates. The derivation of these estimates may range from firm supplier or contractor quotations to establishing costs from concepts alone. In any event, the project costs that are established during the planning phase are those generally used for budget purposes. Therefore, it is important to use the best possible estimates, as these costs become "sacred." The total of the project work activities represents the variable costs of the project budget.

In many instances there is also a fixed total cost. If during the planning phase, the project costs are more than what was allowed in the budget, the costs are reviewed again and revised until they meet the budget objectives.

PROJECT COST SCHEDULE

The principal reasons for scheduling costs are the following:

1. To assist the financial group in funding the project. By projecting its cash flow through a cost schedule, finance personnel will know when and how much of the funds will be needed at any period of time.
2. To assist the project management team in developing the expenditure distribution over the life of the project. Once the project is under way, this planned distribution can be compared with actual costs.
3. The financial activity will also use the cost schedule for property tax purposes and depreciation schedules.
4. Project cost schedules become the basis for determining timing for various property tax payments.

The procedure for developing a cost schedule is as follows:

Step 1. Complete the bar chart time schedule.

Step 2. Prepare a tabulation format to determine the cost slope (the cost/time unit).

Step 3. Determine the time increments at which costs are to be considered. (This is usually one month. Shorter periods of time imply that costs are critical; longer periods suggest that costs are not critical.)

Step 4. Prepare a tabulation format to determine costs (or expenditures) for the time increments.

Step 5. Total the expenditures for each time increment.

Step 6. Total the expenditures for all the time increments, which represents the total project cost.

Step 7. Plot the total for each time increment on a cost distribution graph using the the bar chart time schedule.

Adjustments to the timing schedule may be necessary after an analysis of the cost schedule is made. The procedure described above has a number of advantages that will allow a good cost analysis to be made.

Once the bar chart time schedule is completed, the next step in developing a cost schedule is to prepare the cost slope for each project job. The *cost slope* is equal to the cost incurred in performing a job activity per unit length of time. A suggested format to prepare the cost slope is shown as follows:

Cost Slope

Activity	Total Cost (dollars)	Duration (weeks)	Cost Slope (dollars/week)
(1)	(2)	(3)	(4)

Specific Instructions Column	Instructions
(1) Activity	Enter the description of the project activity.
(2) Total Cost	Enter the total cost in dollars of performing the project activity.
(3) Duration	Enter the total duration time of the project activity. The unit used will be determined by the requirements of the project. Usually, one-month increments are adequate; however, a smaller unit of time is used if the project costs become critical.
(4) Cost Slope	Enter the result of dividing column (2) by column (3).

Once the cost slope is calculated, the next step is to determine the total amount of expenditures that the project incurs per unit of time (week or month is a common timing base). The necessary prerequisites for this step (as in the case of calculating the cost slope) is to have completed the network analysis, since the network diagram is used as a basis to prepare the bar chart time schedule. This schedule is the important base document that is required to be used in conjunction with the total monthly (or weekly) project expenditures. A suggested format for project expenditures is as follows:

Cost Schedule

Time Period	Activity	Activity Time (weeks)	Cost Slope (dollars/week)	Expenditures (dollars)
(5)	(6)	(7)	(8)	(9)

Specific Instructions Column	Instructions
(5) Time Period	Enter the period of time that is desired to determine the expenditures. (Usually, one-month periods are used; however, if project costs appear critical, two-week periods may be required. In any event, the time period should remain constant.)
(6) Activity	Enter the description of the project activity.
(7) Activity Time	Enter the time units that the activity performs within the time period.
(8) Cost Slope	Enter the cost slope that was developed previously.
(9) Expenditures	Enter the result of multiplying column (7) by column (8). Total the expenditures within each time period, and enter the total for each period before starting the next time period.

The total cost of the project will equal the sum of the total time period expenditures. The expenditures developed from this cost schedule format will serve as the basis for calculating the variance from actual project expenditures during a given time period. As an additional expedient the bar chart schedule and the cost distribution graph can be used to develop a graphic portrayal of the project costs for each time period.

Adjustments to the timing schedule may be necessary after making an analysis of the cost schedule and the cost distribution graph. Changes of this type during the planning period are most beneficial, as many problems associated with time/cost relationships can be resolved before the project gets under way. Spending modifications after the project starts may also require scheduling changes, and the bar chart cost schedule with the cost distribution graph will be helpful in the replanning of the project. When scheduling changes due to project costs are required, the first approach is to examine the activities having float times so that the project duration time is not affected.

The sample problem, "A New Product Introduction," will be used to illustrate the procedure in setting up a cost schedule.

1. Finding the cost slope. The cost estimate for each project item had been determined previously and the total project budget is the sum of the project item costs. The first step is the development of the cost slope, which in this problem will be the weekly costs for each project item. (Development of the cost slope is shown in Figure 5-1.)

Project activity	Cost (dollars)	Duration (weeks)	Cost slope (dollars/week)
Design package	$ 7,500	2	$ 3,750
Order stock	2,000	13	154
Organize sales office	12,000	6	2,000
Set up packaging facility	45,000	10	4,500
Package stock	6,000	6	1,000
Hire sales personnel	8,000	4	2,000
Select distributors	13,000	9	1,444
Select advertising agency	4,000	2	2,000
Train sales personnel	28,000	7	4,000
Sell to distributors	64,000	6	10,666
Ship stock to distributors	9,000	6	1,500
Plan advertising campaign	6,000	4	1,500
Conduct advertising campaign	36,000	10	3,600
Total	$240,500		

Figure 5-1   Calculating the cost slope (cost/week).

2. Developing the cost schedule.   Totaling the expenditures for each 5 weeks of the project was selected as the most appropriate for this project. The initial cost schedule for the product introduction project shows the following cash flow for each 5-week period:

Weeks	Total Cost
0–5	$31,770
6–10	46,050
11–15	52,484
16–20	61,998
21–25	42,198
26–29	6,000

Development of the foregoing expenditures for each 5-week period is shown in Figure 5-2. The basis of the cost schedule in the bar chart is shown in Figure 5-1.

3. Constructing the cost distribution graph (from the bar chart time schedule).   The bar chart time schedule and the 5-week cost distribution graph are shown in Figure 5-3.

Period	Activity	Activity time (weeks)	Cost slope (dollars/week)	Total activity expenditures (dollars)
0–5	Design package	2	$ 3,750	$ 7,500
	Order stock	5	154	770
	Set up packaging facility	3	4,500	13,500
	Organize sales office	5	2,000	10,000
	Total 0–5			$31,770
6–10	Order stock	5	154	770
	Set up packaging facility	5	4,500	22,500
	Organize sales office	1	2,000	2,000
	Hire sales personnel	4	2,000	8,000
	Select distributors	4	1,444	5,776
	Select advertising agency	2	2,000	4,000
	Plan advertising campaign	2	1,500	3,000
	Total 6–10			$46,046
11–15	Order stock	3	154	462
	Set up packaging facility	2	4,500	9,000
	Package stock	2	1,000	2,000
	Train sales personnel	5	4,000	20,000
	Select distributors	5	1,444	7,220
	Plan advertising campaign	2	1,500	3,000
	Conduct advertising campaign	3	3,600	10,800
	Total 11–15			$52,482
16–20	Package stock	4	1,000	4,000
	Train sales personnel	2	4,000	8,000
	Sell to distributors	3	10,666	31,998
	Conduct advertising campaign	5	3,600	18,000
	Total 16–20			$61,998
21–25	Sell to distributors	3	10,666	31,998
	Ship to distributors	2	1,500	3,000
	Conduct advertising campaign	2	3,600	7,200
	Total 21–25			$42,198
26–29	Ship to distributors	4	1,500	6,000
	TOTAL PROJECT COSTS			$240,500

Figure 5-2 Calculating the cost schedule.

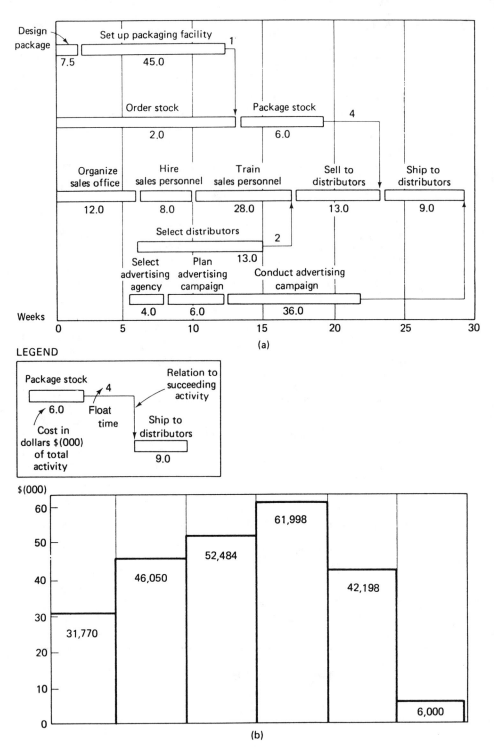

Figure 5-3 (a) Bar chart cost schedule (total cost for each work activity is noted below the corresponding bar); (b) cost distribution graph.

The time schedule for the activities is based on the earliest start times. For making adjustments to the schedule, the free float is shown on the bar chart, thus allowing changes to be made on the chart within the planned project duration.

4. Making schedule adjustments to allow more uniform cost distribution (see Figure 5-4).   It appears that disbursements in the 16–20-week period are appreciably higher than in the other cost periods, and reducing the cash flow in this period may be important to the financial arrangements made for this project. For example, credit procurement may be premised on more even monthly disbursements.

Examining this project reveals that the advertising sequence (or chain), which has high unit costs, has a 7-week float. Pushing "Conducting the Advertising Campaign" to its latest start not only helps in arriving at a more uniform cash flow, but permits the campaign to be under way as close to the end of the project as possible.

Weeks	Total Cost
0–5	$31,770
6–10	46,050
11–15	52,484
16–20	47,598
21–25	53,998
26–29	20,440

Conduct advertising campaign (use latest start)

Period (weeks)	Activity	Actual time (weeks)	Cost slope (dollars/week)	Expenditures (dollars)
16–20	Package stock	4	$ 1,000	$ 4,000
	Train sales personnel	2	4,000	8,000
	Sell to distributors	3	10,666	31,998
	Conduct advertising campaign	1	3,600	3,600
				$47,598
21–25	Sell to distributors	3	10,666	31,998
	Ship to distributors	2	1,500	3,000
	Conduct advertising campaign	5	3,600	18,000
				$53,998
26–29	Ship to distributors	4	1,500	6,000
	Conduct advertising campaign	4	3,600	14,400
				$20,400

Figure 5-4 Revised project costs.

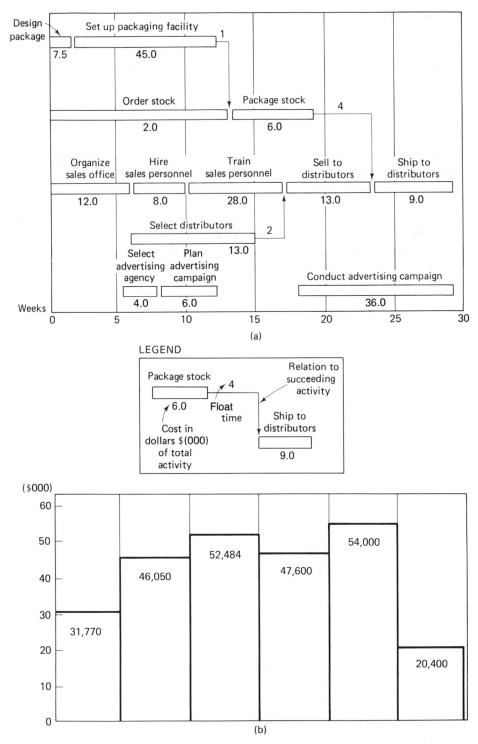

Figure 5-5 (a) Revised bar chart cost schedule; (b) revised cost distribution graph.

5. Constructing the revised bar chart schedule and revised cost distribution graph   The revised bar chart time schedule and the graphic display of the expenditures reflecting the adjusting of the advertising campaign schedule are shown in Figure 5-5.

PROJECT COST CONTROL

In the same manner as the project timing schedule, the status of project costs will be required periodically as the project is under way. As a management by exception application, a procedure should be established that can predict overspending (or cost overruns) as soon as they can be discovered.

The project cost status report or indicated cost outcome report is a control device that is designed to ensure that project spending is contained within approved (authorized) amounts. Specifically, the report is used to review and evaluate this spending status of projects; to determine if project commitments are in line with authorized amounts; and to determine if (and when) additional authorizations may be required.

Usually, someone with a financial background will be responsible for preparing, reviewing, and evaluating these reports and for ensuring that the projected (indicated) costs are realistic. The project management team needs to provide the data when a potential overrun (or underrun) is first disclosed; at that time, a special review and analysis of open commitments and uncompleted portions of the work should be made. If it is determined that the overrun (or underrun) is likely to be beyond allowable tolerances, a request should be submitted for a project supplement as soon as possible. In the case of a project overrun beyond tolerance, no further commitments should be made against the project until additional authorized funding is obtained. It is possible that authorized funding cannot be obtained until those responsible for the project can assure the financial authorities that the total project costs can be contained. If no such assurance occurs, the project could be shut down while a complete project cost analysis is made. Although this may sound drastic, it may be a better approach than prolonging the cost overruns until the project goes bankrupt and there are no more funds to continue the project.

The submission to management of project status reports including the financial activity should be done on a regular basis. A good status reporting method is the *Indicated Cost Outcome Report*. A suggested format follows (p. 4).

If a project is near the end of completion, overruns close to the 10% limit may be tolerable, but if the program is in the early stages, a more in-depth review is necessary. In many firms it may be the perogative of the controller to refuse to authorize any further spending as a project reaches an overrun stage. It is good practice to review the program at periodic intervals with top

Indicated Cost Outcome Report

Date: _____

Project Start Date: _____

Project Completion Date: _____

Project Item	Authorized (Budgeted)	Committed to Date	Future Commitments
(1)	(2)	(3)	(4)

Indicated Outcome	Variance (Over) or (Under)	Percent (Over) or (Under)
(5)	(6)	(7)

Column	Instructions
(1) Project Item	Enter the description with each project job activity. (There may be several items included with some activities. In this case, they should also be itemized.)
(2) Authorized Budget	Enter the estimated dollar amount that has been developed and approved for completing the project activity item. There may be firm quotations that make up this item. If so, the estimate may be satisfactory to list without any added funds or contingency.
	A word on contingency: A general rule that should be followed is: never estimate an item that cannot be reported. This applies to time estimates as well. Contingency is an item that cannot be reported. If there are uncertainties, the project item includes a predetermined added amount within the estimate. Your own internal detail must include this as a separate item. It could be noted as potential added work items, potential design changes, and so on.
	The total costs for each of the project items may also be termed as a budget. The sum of the authorized costs equals the project budget. In most cases the project budget is not compatible with the manner in which an accounting procedure is usually structured. In a project detail each item may have certain account numbers assigned to it and they are used for various purposes. The project costs become the permanent capital investment base. They are also used for tax purposes and such expense items as depreciation. Other accounts may be associated with various operating expense accounts.
(3) Committed to Date	Enter for each activity the amount that has been spent and/or ordered as of the date of the report. When a major cost commitment is made, such as the time an award order is given to a major equipment supplier, an indicated cost outcome should be prepared.
(4) Future Commitments	Enter the additional costs that will be needed to complete the activity. These projections are usually estimated.

Column	Instructions
(5) Indicated Outcome	Enter the sum of column (3), *Committed to Date,* and column (4), *Future Commitments,* for each activity. This figure is compared to the authorized (or budgeted) amount to determine the cost performance of each item. All of the indicated outcomes of the activities are added algebraically, and this total is compared with the total budgeted cost.
(6) Variance	Enter the difference between column (2), *Authorized* amount, and column (5), *Indicated Outcome.* When the difference is over the authorized amount (known as the overrun) beyond a predescribed tolerance (usually 10%) of the authorized amount, project operations may be obligated to stop until there is assurance that no further overruns can be expected.
(7) Percent of Variance Over or Under (optional)	Enter the result of dividing the difference between column (2), *Authorized* amount, and column (5), *Indicated Outcome,* divided by column (2), *Authorized* amount. (Multiply by 100 to arrive at percent amounts.)

management to assure them that the program costs will be contained (or the strategy to be employed where program costs may become a problem).

SAMPLE PROBLEM: A NEW PRODUCT INTRODUCTION

The sample problem will be used to illustrate the procedure in setting up an indicated cost outcome report. For this example we have assumed that the project is in its fifteenth week and a tabulation of the project item costs to date are shown in Figure 5-6. With the indicated cost outcome tabulation, the cost report would include a highlight section as shown in Figure 5-7. In this particular example the program cost overrun may be more than can be tolerated at this period. Unless program costs are shown to be contained, the comptroller may elect to refuse authorization of any additional expenditures. This, in effect, will shut down the project.

Although this may seem harsh, the results would be more disastrous if these costs conditions were not acted upon as early as possible. Shutting down the project is not necessarily an arbitrary decision by the comptroller. There may not be any more funds available for the project other than the authorized amount. The indicated outcome report allows the comptroller to make decisions before the program costs are completely out of control.

Another popular cost status report is an accumulative cost report in the chart in Figure 5-8, showing the total project cost status to date as well as the projected costs. It provides a manually plotted graphic picture of the existing costs compared to the plan as well as the anticipated cost income. Figure 5-9 shows a chart that is generated from the data that have been entered on the computerized spreadsheets.

					Reporting date: Week 15	
					Start date: 0	
					Completion date: Week 29	
Project item	Authorized (budgeted)	Committed to date	Future commitments	Indicated outcome	Variance (over) or under	% Variance (over) or under
Design package	$ 7,500	$ 8,000	$ —	$ 8,000	$(500)	(6.7)
Order stock	2,000	2,000	—	2,000	—	—
Organize sales office	12,000	10,500	—	10,500	1,500	8.3
Set up packaging facility	45,000	50,000	—	50,000	(5,000)	(11.1)
Package stock	6,000	1,500	3,000	4,500	1,500	25.0
Hire sales personnel	8,000	4,000	3,500	7,500	500	6.3
Select distributors	13,000	12,000	—	12,000	1,000	7.7
Select advertising agency	4,000	2,000	—	2,000	2,000	50.0
Train sales personnel	28,000	25,000	7,000	32,000	(4,000)	(14.3)
Sell to distributors	64,000	1,000	63,000	64,000	—	—
Ship to distributors	9,000	—	9,000	9,000	—	—
Plan advertising campaign	6,000	5,500	—	5,500	500	8.3
Conduct advertising campaign	36,000	10,000	36,000	46,000	(10,000)	(27.7)
Total	$240,500	$131,500	$121,500	$253,000	$(12,500)	(5.2)

LEGEND

Authorized — estimated dollar amount.
Committed to date — amount spent or committed to be spent to date.
Future commitments — funds still needed to complete activity.
Indicated outcome — sum of funds committed to date and future commitments.
Variance (over) or under — difference between authorized funds and indicated outcome funds.
Percent (over) or under — [variance divided by authorized funds] times 100.

Figure 5-6   Indicated cost outcome report (week 15).

COST MINIMIZING

Cost minimizing as related to network analysis is concerned with determining how to reduce the time required for completing a project with the least amount of added expense. Reducing the duration of a project would include the review of such items as overtime, extra personnel, and additional equipment. The cost-minimizing technique may also be termed time/cost trade-offs or "crash" programs.

   If it is decided that the project duration time is to be reduced, it will be necessary to obtain the following cost data for each project item:

1. The expenditures required for accomplishing the work on a "normal" time basis.
2. The expenditures required for accomplishing the work on an expedited or "crash" basis. (It will also be necessary to obtain the reduction in time that served as a basis for these expenditures.)

The two sets of estimates will then be used to develop alternative schedules and, from these options, determine the best job schedule in terms of minimum additional cost.

Reporting date: Week 15

- Project expenditures are 5.2% over budget.

- Commitments to date: 55% of total authorized.

- Project is 50% complete.

- Outstanding items that indicate high overruns.
  a. Train sales personnel   (14.3%)
  b. Advertising campaign   (27.7%)

Recommendations to reduce overruns:

a. Train sales personnel — projected expenditures, $7,000; total indicated cost, $32,000; variance, $4,000 over authorized amount.

   Recommendation: Modify training program to permit sales personnel to complete course earlier. Reducing time of training program could reduce project costs by $3,000.

b. Conduct advertising campaign — projected expenditures, $36,000; total indicated cost, $36,000; variance $20,000 over authorized amount.

   Recommendation: Revise TV commercials, reduce advertisements in trade periodicals for an $8,000 savings.

Detailed report on reducing overrun will be included in the next report.

Figure 5-7 Indicated cost outcome—highlights.

The planning and scheduling functions are carried out for the project in the same manner as described in earlier chapters. The cost minimizing technique uses network analysis to develop the alternative schedules. Reviewing the steps in network analysis: first, state the objectives; second, determine the job activities and their interrelationships; third, prepare an arrow diagram; and finally, from the arrow diagram and the designated time estimates for each activity, calculate the timing for the initial schedule.

For cost-minimizing purposes, the following steps are taken:

1. Determine how much time each job can be reduced by "crashing" each job in the project.
2. Obtain the cost for accelerating the work.

From these data, the extra cost that will be incurred can be determined for reduced project duration times until the project is fully crashed. Also, from these data, the optimum project duration in terms of total cost can be deter-

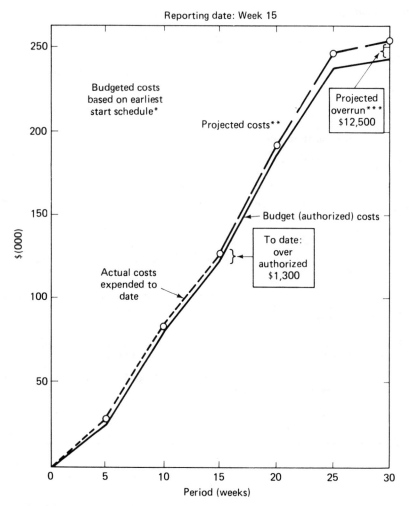

Reporting date: Week 15

Budgeted costs
based on earliest
start schedule*

Projected costs**

Projected
overrun***
$12,500

Budget (authorized) costs

To date:
over
authorized
$1,300

Actual costs
expended to
date

$(000)

Period (weeks)

*See Figure 5-2.
**Projected costs—total funds needed to complete project.
***Projected overrun—additional funds required to complete project.

LEGEND

| Period | Budgeted costs | |
(weeks)	Cost	Accumulated cost
0 - 5	$31,770	$ —
6 - 10	46,050	77,820
11 - 15	52,484	130,304
16 - 20	61,998	192,302
21 - 25	42,198	234,500
26 - 29	6,000	240,500

Figure 5-8   Project costs status (accumulation of costs, actual and projected).

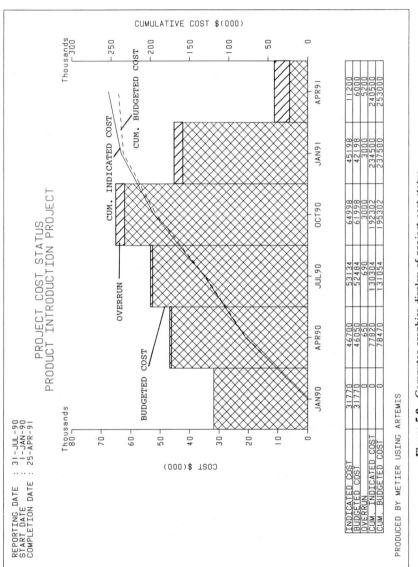

Figure 5-9 Computer graphics display of project cost status.

mined, that is, the minimum amount of additional costs to achieve the best reduced project time.

Procedure

The cost minimizing procedure follows a "cut-and-try" pattern. For a large project, a computer is used to perform the calculations with great speed. However, the work can be done manually and the method shown is a manual arrangement. Once the arrow diagram has been developed, time estimates obtained, and the initial schedule calculated, the following steps are taken:

Step 1. *Direct Cost*
a. Obtain the normal cost and the crash time and crash cost for each job in the network. Total the normal costs for each job to obtain the normal cost for the project.
b. Determine the minimum cost for reducing the project duration by one time interval (such as a day or week). This involves cutting back the duration of those jobs on the critical path that can be reduced at least expense. If more than one path is critical, this procedure is applied to all such paths.
c. Perform the same process to reduce the project duration a second time interval.
d. Repeat the process to a point where the project is "fully crashed" in terms of critical jobs.

Step 2. *Indirect Cost*
a. Determine the indirect cost for the project for the normal and crash times and for the time intervals between them.

Step 3. *Total Cost*
a. Add the direct to the indirect costs to determine the total cost at the various time intervals considered.
b. Identify the time interval, or project duration, at which the total cost will be at a minimum.

Network analysis is based on the premise that time and cost are interrelated. Most projects can be performed in a number of different ways from minimum cost/maximum time to minimum time/maximum cost. The method permits an educated choice between the two extremes, a choice that will be best for the particular operation under consideration.

Method for Calculating Cost Slope

The cost slope gives the rate of increase in cost for the decrease in time. To calculate the cost slope for each activity, four values are required for each activity:

1. *Normal time:* job time estimate that assumes employment of the usual amount of labor, equipment, and so on.
2. *Normal cost:* estimated expense for performing the project within the normal time estimate.
3. *Crash time:* minimum estimated time in which a job could be completed if the job is accelerated by using one or more factors, such as overtime, extra labor, or additional equipment.
4. *Crash cost:* normal cost plus the extra cost involved in applying those factors—overtime, extra labor, additional equipment.

These four factors are used to calculate the cost slope for each job, the increase in cost per unit of reduced project time:

$$\text{Cost slope} = \frac{\text{Crash cost} - \text{Normal cost}}{\text{Normal time} - \text{Crash time}}$$

Certain assumptions must be made in using the cost-minimizing program to accelerate a project:

1. Crash time is always less than or equal to normal time.
2. Crash cost is always greater than or equal to normal cost.
3. The time/cost curve is generally linear, as illustrated in Figure 5-10.

If a more accurate approximation of the time/cost curve is desired, the job may be broken into two or more segments with a linear time/cost approximation for each segment. This is illustrated in Figure 5-11.

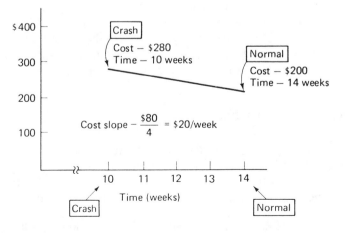

Figure 5-10   Time/cost curve for crashing a job.

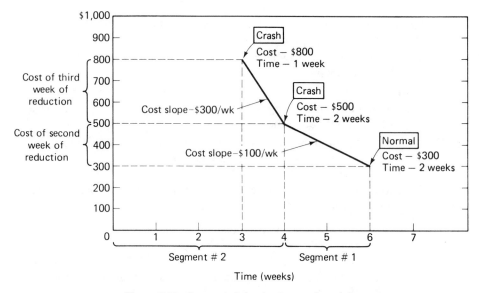

Figure 5-11 Segmented time/cost curve for a job.

Establish the Final Schedule

Determining what the final schedule should be, in terms of reducing the project duration with a minimum overall project cost, can become quite complicated. For example, if the critical path is not much longer in time than other paths, a reduction in time for the critical path may cause one or more other paths to become critical. In many cases there may be a computer program available to provide the calculations more economically. The final schedule can then be established with the assurance that, based on the time estimates and cost data, a minimum project cost will be incurred in meeting the specified deadline.

As a part of the process of establishing the final schedule, the computer is programmed to provide a minimum-cost curve for the project and the schedules corresponding to each point on this curve. Figure 5-12 shows the type of curve that can be printed out for a sample project.

EXAMPLE: OFFICE BUILDING[1]

The following example illustrates the procedure to be followed in applying cost minimizing to a project that is to be accelerated. In this example an industrial plant office facility is to be rebuilt and expanded. The existing office abuts a manufacturing building in a highly congested industrial area. Since no new land

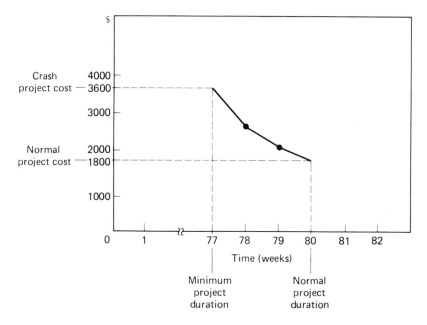

Figure 5-12 Minimum-cost curve for a project.

is available, the existing office building must be demolished before the new building can be constructed in the cleaned area. For simplicity, the office building project has been subdivided into just 12 activities. The network diagram in Figure 5-13 reflects the logical flow of work in the project. Also shown is the tabulation of the normal time schedule of the project stating when each job must be done, when deliveries must take place, which are the critical jobs, and when the project will be completed (in this case, 65 days).

If the 65-day time duration is unacceptable, time can be saved by expediting one or more of the jobs along the critical path. This is more efficient than placing all jobs along the critical path on a crash basis. For each activity that is assigned a normal time duration, there is also a crash time duration—a minimum time in which the activity can be performed. For each of the time durations, there are also associated normal and crash costs. Once the normal crash times and costs are known, the cost slope for each activity can be determined.

$$\text{Cost slope} = \frac{\text{Crash cost} - \text{Normal cost}}{\text{Normal time} - \text{Crash time}}$$

The cost slope gives the rate of increase in cost for decrease in time. The cost data and the cost slope calculations are shown in Figure 5-14.

An intelligent choice of which job to expedite can be made by starting with compressing the time schedule of the critical job with the least-cost slope

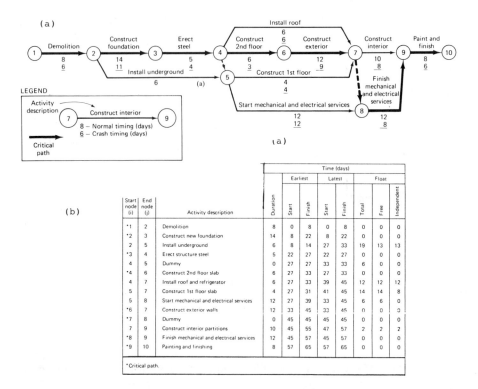

(a)

(b)

(c)

REPORT TYPE : OFFICE SCHEDULE : Normal
TIME NOW DATE: 14/JAN/91   PRODUCED BY: P1211

Node Numbers	Activity Description	Early Start	Late Start	Early Finish	Late Finish	Float
001 -002	DEMOLITION	14/JAN/91	14/JAN/91	23/JAN/91	23/JAN/91	0
002 -003	CONSTRUCT NEW FOUNDATION	24/JAN/91	24/JAN/91	12/FEB/91	12/FEB/91	0
002 -006	INSTALL UNDERGROUND	24/JAN/91	20/FEB/91	31/JAN/91	27/FEB/91	19
003 -004	ERECT STRUCTURE STEEL	13/FEB/91	13/FEB/91	19/FEB/91	19/FEB/91	0
004 -006	CONSTRUCT 2ND FLOOR SLAB	20/FEB/91	20/FEB/91	27/FEB/91	27/FEB/91	0
004 -007	INSTALL ROOF	20/FEB/91	8/MAR/91	27/FEB/91	15/MAR/91	12
005 -007	CONSTRUCT 1ST. FLOOR SLAB	20/FEB/91	12/MAR/91	25/FEB/91	15/MAR/91	14
005 -008	START MECHANICAL AND ELECTRICAL SERVICES	20/FEB/91	28/FEB/91	7/MAR/91	15/MAR/91	6
006 -007	CONSTRUCT EXTERIOR WALLS	28/FEB/91	28/FEB/91	15/MAR/91	15/MAR/91	0
007 -009	CONSTRUCT INTERIOR PARTITIONS	18/MAR/91	20/MAR/91	29/MAR/91	2/APR/91	2
008 -009	FINISH MECHANICAL AND ELECTRICAL SERVICE	18/MAR/91	18/MAR/91	2/APR/91	2/APR/91	0
009 -010	PAINTING AND FINISHING	3/APR/91	3/APR/91	12/APR/91	12/APR/91	0

Figure 5-13   (a) Network diagram; (b) time schedule (manual); (c) computerized schedule.

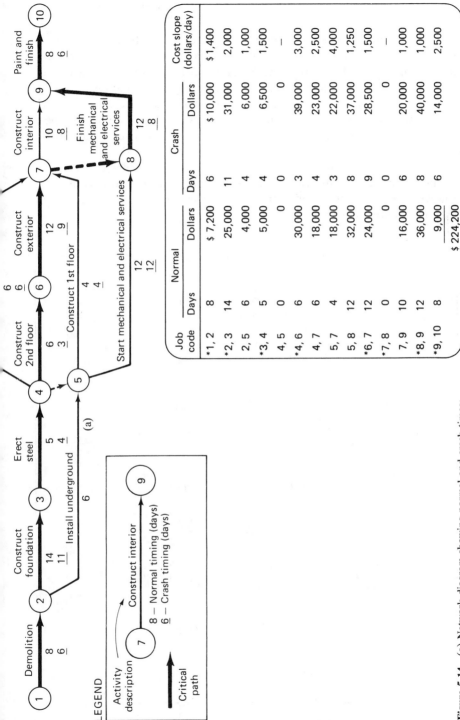

LEGEND

Activity description → Construct interior

8 – Normal timing (days)
6 – Crash timing (days)

Critical path

(a)

Job code	Normal		Crash		Cost slope (dollars/day)
	Days	Dollars	Days	Dollars	
*1, 2	8	$ 7,200	6	$ 10,000	$ 1,400
*2, 3	14	25,000	11	31,000	2,000
2, 5	6	4,000	4	6,000	1,000
*3, 4	5	5,000	4	6,500	1,500
4, 5	0	0	0	0	–
*4, 6	6	30,000	3	39,000	3,000
4, 7	6	18,000	4	23,000	2,500
5, 7	4	18,000	3	22,000	4,000
5, 8	12	32,000	8	37,000	1,250
*6, 7	12	24,000	9	28,500	1,500
*7, 8	0	0	0	0	–
7, 9	10	16,000	6	20,000	1,000
*8, 9	12	36,000	8	40,000	1,000
*9, 10	8	9,000	6	14,000	2,500
		$ 224,200			

*Critical path.

(b)

Figure 5-14 (a) Network diagram showing normal and crash times; (b) normal crash data.

and then compressing the jobs with the lesser-cost slopes. In this example, the least-cost slope exists with Job 8,9, where a 2-day reduction can be affected at an additional cost of $1,000 per day. At this point Job 7,9 also becomes critical and any further reduction in Job 8,9 would also require a reduction in Job 7,9, which would increase the cost by $2,000 per day. Since Job 1,2 can be reduced in duration by 2 days at a cost of $1,400 per day, this job would be the next one to compress. Shown in Figure 5-15, this process is repeated until the project duration has been reduced to its minimum time of 47 days.

Since a decrease in project time requires additional capital expenditures, the direct costs obviously rise as the project time is decreased from 65 days. However, the total cost of this particular project is comprised of both direct costs and indirect costs.

Project duration (days)	Normal cost	Job code	Reduced time (days)	Additional cost	Normal and crash costs
65	$224,200	–	–	–	$224,200
63		8, 9	2	$2,000	226,200
61		1, 2	2	2,800	229,000
59		3, 4	1	1,500	
		6, 7	1	1,500	
			2	3,000	232,000
57		6, 7	2	3,000	235,000
55		2, 3	2	4,000	239,000
53		8, 9	2	2,000	
		7, 9	2	2,000	
			2	4,000	243,000
51		2, 3	1	2,000	
		9, 10	1	2,500	
			2	4,500	247,500
49		9, 10	1	2,500	
		4, 6	1	3,000	
			2	5,500	253,000
47		4, 6	2	6,000	259,000

Figure 5-15 Calculations of direct costs and total costs to crash the Office Building Project.

Indirect costs consist of such items as overhead, insurance, interest on capitalization, production losses, and liquidated damage clauses. (Liquidated damages are paid to the owner if the project is not completed on time. Government contracts usually have this clause included.) Indirect costs have a tendency to decrease in cost with an increase in project duration. Tabulations for indirect costs for this example are shown in Figure 5-16.

In this example the cost slope of the indirect costs ($2,000/day) is greater than the weighted average of cost slopes of the direct costs (ranging from $1,000 to $6,000/day). As a result, the total cost of this project is reduced as the duration time is reduced. The project duration can be crashed from 65 days to 57 days to reach the minimum cost. If time is a premium, the project can be crashed to its minimum duration time of 47 days. As the total project costs of $295,000 to perform work in 47 days is about the same as the budgeted $294,200 to perform work in 65 days, there was no cost penalty to expedite work. If minimum costs are important, the project can be completed in 57 days for $286,000, the lowest total project cost.

Figure 5-17 displays the total cost curve for the 47–65-day period, which shows clearly the optimum project duration time which falls somewhere between the normal time and the crash time when the total project costs (the sum of indirect and direct costs) are considered.

The last step in this procedure is to recalculate the project schedule once the revised job estimates are established for the new project duration time. This

Project duration (days)	Direct costs	Indirect costs Normal	Indirect costs Crash	Indirect costs (normal and crash)	Total cost (direct and indirect)
65	$224,200	$70,000	–	$70,000	$294,200
63	226,200		$(4,000)	66,000	292,200
61	229,000		(4,000)	62,000	291,000
59	232,000		(4,000)	58,000	290,000
57	235,000		(4,000)	54,000	286,000
55	239,000		(4,000)	52,000	291,000
53	243,000		(4,000)	48,000	291,000
51	247,500		(4,000)	44,000	291,500
49	253,000		(4,000)	40,000	293,000
47	259,000		(4,000)	36,000	295,000

*Indirect cost slope: ($2,000 per day) – indirect costs include overhead, insurance, interest on capital loan, etc.

Figure 5-16 Calculations of additional indirect costs and total costs to crash the Office Building Project.

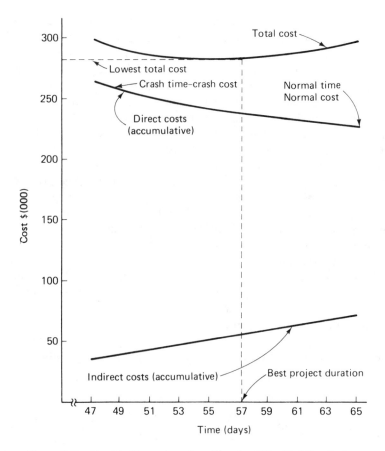

Figure 5-17   Graphic illustration of crashing the Office Building Project.

serves not only as a check but provides a new schedule once the revised job estimates are established for the new project duration time.

Summarizing the cost-minimizing technique: It is used in conjunction with the network planning method to determine the optimum completion time of a project or to accelerate the completion of a project, where necessary, at minimum additional cost. The information provided is of great value to management in the initial scheduling of a project and in making adjustments when there is a delay in accomplishing specific parts of the work.

Two assumptions must be made in using the cost-minimizing procedure: Crash time is always less than or equal to normal time, and crash cost is always greater than or equal to normal cost. The procedure includes determining the direct cost, the indirect cost, and the total cost for the project in terms of the normal time and the crash time and the time intervals in between.

The cost-minimizing method can be applied manually for a small project; however, for a large project, a computer must be used if the work is to be done effectively. With the cost-minimizing computer program, the computer will make all the normal and crash time calculations and compute the minimum-cost curve. It will also print all the data required for scheduling the project, including earliest and latest start, earliest and latest finish, and float.

When used effectively with network planning, the cost-minimizing technique, whether done manually or with a computer program, provides management with an effective tool that can be used in planning and scheduling projects.

SUMMARY

Cost management may be considered the most important element of project management, with the management of time a close second. Whereas time delays may be tolerated to some degree, funds allocated to the project cannot be exceeded or, at best, will require considerable efforts to acquire additional funds. This chapter concerns itself with several fundamental aspects of project cost areas that can be used effectively for project cost control.

- *Cost schedule:* distribution of project costs over the life of the project. Once a project is under way, actual expenditures are compared periodically with the planned costs (also known as budgeted costs). With computerized spreadsheets and project management software packages with cost spreadsheet features, the tedious efforts of manual data and calculations have essentially been eliminated. One type of time/cost reporting now available (Figure 5-18) is used to initiate the countless other computerized reports that can be generated to assist in the cost management of the project.

- *Cost control:* an extension of cost scheduling comparing actual expenditures with budgeted costs on a periodic basis. An application known as indicated cost outcome predicts overspending situations early enough in a project to allow for planning corrections. This early-warning system uses electronic spreadsheets and accompanying software as effective tools.

- *Cost minimizing:* reducing project duration with the least amount of added expense. This method requires that a planning model and initial schedule be completed. Next, critical items are investigated for possible crashing—duration reduced through overtime and additional resources. Specialized software is available for applying cost-minimizing techniques.

REPORT TYPE : PROJECT ITEM TIMING & COSTS : Normal
TIME NOW DATE: 1/APR/90   PRODUCED BY: P1211

Node Numbers	Description	CT	Early Start	Early Finish
034 -002	DESIGN PACKAGE	7500	2/APR/90	13/APR/90
002 -003	SET UP PACKAGING FACILITY	45000	16/APR/90	22/JUN/90
001 -003	ORDER STOCK	2015	2/APR/90	29/JUN/90
003 -007	PACKAGE STOCK	6000	2/JUL/90	10/AUG/90
007 -011	SHIP STOCK TO DISTRIBUTORS	9000	10/SEP/90	19/OCT/90
001 -004	ORGANIZE SALES OFFICE	12000	2/APR/90	11/MAY/90
004 -008	SELECT ADVERTISING AGENCY	4000	14/MAY/90	25/MAY/90
004 -005	HIRE SALES PERSONNEL	8000	14/MAY/90	8/JUN/90
008 -009	PLAN ADVERTISING CAMPAIGN	6000	28/MAY/90	22/JUN/90
005 -006	TRAIN SALES PERSONNEL	28000	11/JUN/90	27/JUL/90
009 -010	CONDUCT ADVERTISING CAMPAIGN	36000	25/JUN/90	31/AUG/90
004 -006	SELECT DISTRIBUTORS	13005	14/MAY/90	13/JUL/90
006 -007	SELL TO DISTRIBUTORS	63990	30/JUL/90	7/SEP/90

(a)

Node Numbers	Description	CT	Late Start	Late Finish
034 -002	DESIGN PACKAGE	7500	7/MAY/90	18/MAY/90
002 -003	SET UP PACKAGING FACILITY	45000	21/MAY/90	27/JUL/90
001 -003	ORDER STOCK	2015	30/APR/90	27/JUL/90
003 -007	PACKAGE STOCK	6000	30/JUL/90	7/SEP/90
007 -011	SHIP STOCK TO DISTRIBUTORS	9000	10/SEP/90	19/OCT/90
001 -004	ORGANIZE SALES OFFICE	12000	2/APR/90	11/MAY/90
004 -008	SELECT ADVERTISING AGENCY	4000	2/JUL/90	13/JUL/90
004 -005	HIRE SALES PERSONNEL	8000	14/MAY/90	8/JUN/90
008 -009	PLAN ADVERTISING CAMPAIGN	6000	16/JUL/90	10/AUG/90
005 -006	TRAIN SALES PERSONNEL	28000	11/JUN/90	27/JUL/90
009 -010	CONDUCT ADVERTISING CAMPAIGN	36000	13/AUG/90	19/OCT/90
004 -006	SELECT DISTRIBUTORS	13005	28/MAY/90	27/JUL/90
006 -007	SELL TO DISTRIBUTORS	63990	30/JUL/90	7/SEP/90

(b)

Figure 5-18   (a) Project activity timing and costs for the early start schedule; (b) late start schedule.

6
Resource Leveling

We have shown so far that through the proper use of the network analysis technique we can plan, schedule, and control projects with some degree of confidence. However, in following the project management cycle to this point, we have made one major assumption:

> There is no restriction on availability of labor,
> equipment, and/or other resources

In applications where labor or equipment limitations are not a restriction, the network planning technique can adequately deal with the planning, scheduling, and controlling of projects. However, in most situations there are labor and equipment shortages, and although enough labor may be available to complete the critical jobs on time, there may not always be enough workers to do all noncritical jobs in the time specified. Since the resource considerations are not inherent in network planning, but are necessary for a proper planning and scheduling effort, the resource-leveling technique is used to produce a practical solution to the problem.

This is not a simple technique to apply. Planning the efficient use of resources is a complex task. One must not only effectively allocate resources among projects to meet scheduled target dates, but also select the most effective technical approach to the engineering and production work required for each project.

It is not unusual to have labor and equipment shortages. Physical restrictions, although not usually considered in the same light as availability limitations, can also be a major obstacle to straight-forward scheduling. An example of a physical restriction is a confined location where only one person can work at a time; another example is a job that can be completed more safely or efficiently in daylight. Some of these limitations can be overcome if they are recognized in time. Where additional resources are needed, equipment can be rented, work may be contracted out, and for some projects, additional workers can be used on a part-time basis. All of these alternatives add to the cost of the project. Therefore, it will be the planner's objective to minimize these additional costs.

Several definitions for resources are used in applying resource leveling. For our purposes, resources are limited to labor, equipment, facilities, and financial budgets. All of these factors must be considered when analyzing the duration of a project.

As labor is the main resource in most projects, our discussion will be directed to allocating labor. Our objective will be to apply the available resources within the prescribed project time limits. To achieve maximum efficiency within these parameters, an organization strives for the following goals:

1. Reduce the peaks and valleys in labor demands.
2. Minimize crew size.
3. Avoid idle or downtime.
4. Balance the overall labor requirements over reasonable periods of time.

Resource leveling also means resource stabilizing. Stabilizing the work force and achieving the goals listed above have the following advantages:

1. The main objective of any industry should be a consistent-project-life employee. Almost every organization wants employees to feel that once hired, they can be reasonably sure of consistent employment.
2. An organization wants to avoid layoffs and rehires. Not only is bringing workers in and out of a project costly and inefficient, but the peak demands are often difficult to meet.
3. There is another intangible inefficiency—workers "smell" the end of a job much better than supervisors can at times, and a subtle slowdown begins.
4. The situations described can exist not only in construction and industrial plants, but also in the technical and professional engineering fields.

The difficult part of solving the problem of resource leveling in the mathematical sense is usually the lack of explicit criteria with which one can obtain the best use of resources. To establish a base, it will be necessary to arbitrarily establish available levels of resources as well as changes in levels. This situation exists to some degree in most of the organizations involved.

Techniques of Resource Leveling

In scheduling labor, whether for a large or a small project, to complete the project on time, the highest priority for scheduling orders are the jobs on the critical path. Changing the scheduling of the critical jobs to adjust for leveling out labor is the last process. Therefore, the sequence to use in scheduling is as follows:

- Job with the least float
- Remaining noncritical jobs

In small jobs a manual leveling procedure can be used. When the project is large, computer programs are available that follow the same procedure as the manual approach to assist resource-leveling purposes. Use of the computer will be discussed later in this chapter.

RESOURCE-LEVELING PROCEDURES

This procedure can be used after: first, defining the objectives; second, preparing a network diagram; and then, after preparing a schedule of the project, defining the critical and noncritical work items. The following steps should be completed:

Step 1. Plot all work items on a bar chart by early start times.

Step 2. Schedule all work items to start at their earliest start time if all required resources are available.

Step 3. If all required resources are not available, delay the start of each item until the resources are available within the float time of the specific work item.

Step 4. Make necessary scheduling adjustments in the following order:
- Noncritical jobs
- Jobs almost on the critical list
- Critical jobs

Step 5. If the start of a work item is delayed so much that it cannot be completed by its latest finish time and available float is exceeded, there are two options to level the available labor:
- Increase the resource limit for that activity.
- Schedule the start of the work items as soon as resources permit.

Step 6. If you cannot increase the resource limit and the start of work is to be delayed, you need to examine the effect on all work items dependent on it.

Step 7. When recomputing the network schedule, you must keep fixed the start and finish times of the work items already scheduled.

Step 8. Continue the process until all work items are scheduled.

RESOURCE ALLOCATION ILLUSTRATION: BUILDING DESIGN PROJECT[1]

To illustrate the manual method for labor scheduling, we consider the plan of an architectural engineering firm that is preparing a set of building designs and specifications.

Project Objectives

1. Complete the project in 8 weeks instead of the 9 weeks that were originally scheduled.
2. Maintain a constant crew size of six engineers.

From the network planning diagram in Figure 6-1, we know that Jobs A, D, F, and G are on the critical path, and since there is no float time, these jobs (with the possible exception of A) cannot change from a timing or labor standpoint. Therefore, our leveling tasks will deal with Jobs B, C, and E, which have float or optional starting and finishing times.

Our initial leveling efforts will follow this procedure:

- Draw a simple bar chart showing each job starting at its earliest start time and continuing for its assigned duration.
- Show the weekly (or whatever unit of time is being considered) crew on the bar chart (see Figure 6-2).
- Total the daily crew. In using the earliest start as a basis for the schedule, there is a variation in assignments, ranging from a maximum of eight engineers down to one engineer over the span of 9 weeks. This schedule also reflects poor work continuity, and the project completion time remains 9 weeks, which indicates that neither project objective has been used.
- By graphing the daily crew requirements, one can readily see the peaks and valleys of using the earliest start schedule (see Figure 6-3).

[1]Adapted from Byron M. Radcliffe, Donald E. Kowal, and Ralph J. Stephenson, *Critical Path Textbook* (Chicago: Cahners Publishing Company, Inc., 1967), pp. 101–104.

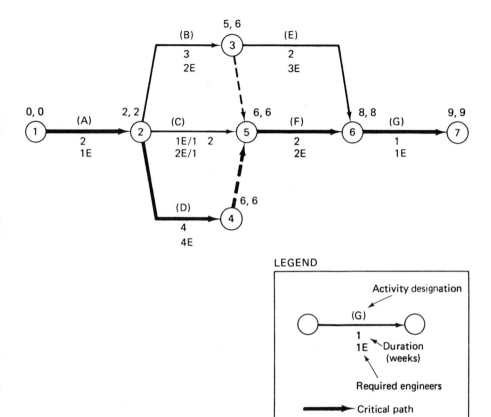

Job	Description	Latest finish	Earliest start	Duration (weeks)	Total float
1, 2 (A)	Preliminary design	2	0	2	0*
2, 3 (B)	Specification	6	2	3	1
2, 4 (D)	Building design	6	2	4	0*
2, 5 (C)	Site design	6	2	2	2
3, 6 (E)	Specification and proposal	8	5	2	1
5, 6 (F)	Review	8	6	2	0*
6, 7 (G)	Submit design and specification package	9	8	1	0*
4, 5	Dummy	6	6	0	0*
3, 5	Dummy	6	5	0	1

*Critical path items.

Figure 6-1 (a) Network diagram; (b) total float tabulations.

Job	Working weeks 1	2	3	4	5	6	7	8	9	Total engineering weeks
A*	—1E—┼—1E—									2
B			2E _ 2E __ 2E →							6
C			_1E _ 2E _ →							3
D*			4E _ 4E __ 4E __ 4E							16
E						3E _ 3E →				6
F*	LEGEND Float ⬚⬚⬚→ 2E						2E __ 2E			4
G*	Required engineers								1E	1
Total engineers	1	1	7	8	6	7	5	2	1	38

* Critical path.

Figure 6-2 Labor leveling: bar chart time schedule (using earliest start time).

Drawing the chart by the graphical method is fairly simple. The horizontal scale represents the weeks (or whatever unit of time is being used), and the vertical scale represents the number of engineers (or whatever type of labor or skills that would be used).

Each work item is drawn as a rectangle whose length represents the duration of the activity and the height is the number of personnel required for that job. The duration is taken from the bar chart.

Showing both the bar chart schedule and the graphical representation on the same sheet has some decided advantages. It affords the opportunity to show the effect on the labor graph at the same time that adjustments are made on the bar chart time schedule.

Additional leveling efforts are required, as there exists poor work continuity and the work schedule has not been shortened. The next move is to reassign some of the float time from the noncritical jobs.

- The first step is a fairly simple approach. Adjust the bar chart time schedules to show the noncritical jobs starting at their *latest start* and continuing for its assigned duration.
- With all noncritical jobs starting at their latest start, check work force continuity and make the necessary adjustments.
- Through further adjustments within the range of the optional starting and

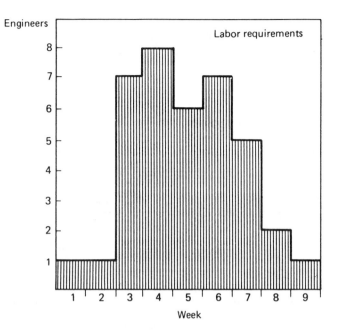

Figure 6-3 Graphic load chart of labor allocation (using earliest start time).

finishing times, the time schedule is arranged. (There is one week during which the crew size will be more than the objective crew size of six engineers. If it is necessary to hold the six-engineer crew size, overtime or extending the project will be required. As the project date is apparently fixed, overtime for one or several work activities appears to be necessary.)

• To investigate the possibility of reducing the project duration, observe the jobs that can be reduced in time by doubling the crew size. In this case we were able to reduce the time of Job A by 1 day by doubling the number of workers. (In many cases, doubling the crew size would not necessarily cut the required time in half.)

• The last step is to review the network diagram to see if the schedule changes made necessitate a change in the network plan.

Figures 6-4 and 6-5 show the final schedule and graphical representation of leveling the labor in this example.

This example is rather simple, and the analysis, if followed in the sequence that has been outlined, is not too difficult. This procedure can also be used for multiple resources; however, scheduling and resource allocation can become complex with larger and more involved projects. When this occurs, we resort to the computer. The use of the computer for assisting in resolving resource-leveling situations is described in the remaining portion of this chapter.

Job	Working weeks								Total engineering weeks
	1	2	3	4	5	6	7	8	
A	2E								2
B		2E	2E		2E				6
C				2E	1E				3
D		4E	4E	4E	4E				16
E						3E	3E		6
F						2E	2E		4
G								1E	1
Total engineers	2	6	6	6	7	5	5	1	38

LEGEND

2E Required engineers

Figure 6-4 Final labor leveling: adjusted schedule.

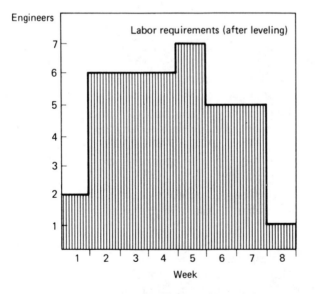

Figure 6-5 Final graphic load chart.

RESOURCE LEVELING BY COMPUTER

Project management software can handle the countless numbers of arithmetical and logical calculations needed to make the resource-leveling decisions. Computer programs can support automatic resource leveling by "smoothing" project activities within the designated completion date or by scheduling a revised completion date based on the resource availability. Therefore, if desired, this feature can allow requirements to control the length of the project.

As all of these computer programs featuring leveling use the input data from the planning diagrams and the computerized schedule, the project planning and scheduling procedures as discussed in earlier chapters are now applied and are illustrated in the following example.

RESOURCE ALLOCATION EXAMPLE—COMPUTER APPLICATION: BUILDING PROJECT

Preparing a work breakdown structure (WBS) is one of the first steps in the Building Project and will help identify the skilled labor resources needed for this project. The WBS (shown in Figure 6-6), together with the initial plan and

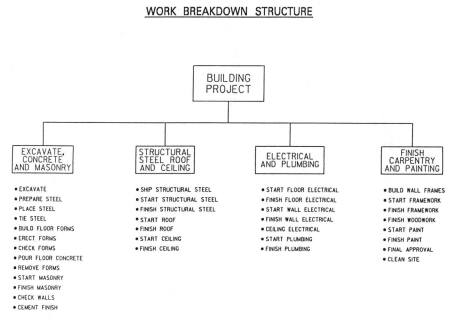

Figure 6-6   Work breakdown structure.

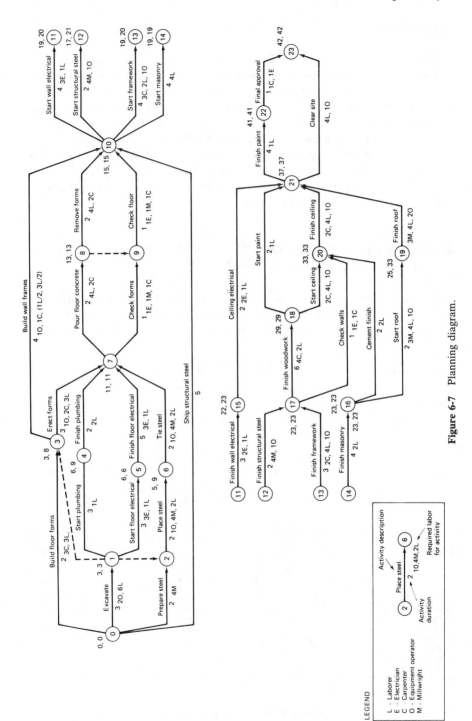

Figure 6-7 Planning diagram.

schedule and the resource planning data, may be submitted to proposed contractors as a guide in their implementation of the Building Project. The Building Project, whose network is shown in Figure 6-7, will be used to illustrate resource leveling utilizing an available computer program.

When the timing data are input into the computer, we can get a report showing the earliest start, latest start, earliest finish, latest finish, and float times for each activity. The computer printout report, with activities in a WBS order, for the sample problem is shown in Figure 6-8. This project starts on April 1, 1991, and is to be completed on January 10, 1992, a period of about 41 weeks. In the initial effort (as we recognize that the effect of the total resources

BUILDING PROJECT

REPORT TYPE : PROJECT SCHEDULE REPORT : Normal
TIME NOW DATE: 1/APR/91   PRODUCED BY: P1211

NODE NUMBERS	Description	EARLY START	LATE START	EARLY FINISH	LATE FINISH	DURATION (DAYS)	FLOAT (DAYS)
000 -001	EXCAVATE	1/APR/91	1/APR/91	19/APR/91	19/APR/91	15	0
000 -002	PREPARE REINFORCING STEEL	1/APR/91	6/MAY/91	12/APR/91	17/MAY/91	10	25
002 -006	PLACE REINFORCING STEEL	22/APR/91	20/MAY/91	3/MAY/91	31/MAY/91	10	20
006 -007	TIE REINFORCING STEEL	6/MAY/91	3/JUN/91	17/MAY/91	14/JUN/91	10	5
000 -003	BUILD FLOOR FORMS	1/APR/91	13/MAY/91	12/APR/91	24/MAY/91	10	15
003 -007	ERECT FORMS	6/MAY/91	27/MAY/91	24/MAY/91	14/JUN/91	15	15
007 -009	CHECK FORMS	17/JUN/91	1/JUL/91	21/JUN/91	5/JUL/91	5	10
007 -008	PLACE FLOOR CONCRETE	17/JUN/91	17/JUN/91	28/JUN/91	28/JUN/91	10	0
017 020	CHECK WALLS	2/SEP/91	4/NOV/91	6/SEP/91	8/NOV/91	5	15
008 -010	REMOVE FORMS	1/JUL/91	1/JUL/91	12/JUL/91	12/JUL/91	10	0
009 -010	CHECK FLOOR	1/JUL/91	8/JUL/91	5/JUL/91	12/JUL/91	5	5
010 -014	START MASONRY	15/JUL/91	2/SEP/91	9/AUG/91	27/SEP/91	20	35
014 -016	FINISH MASONRY	12/AUG/91	30/SEP/91	6/SEP/91	25/OCT/91	20	35
016 -020	CEMENT FINISH	9/SEP/91	28/OCT/91	20/SEP/91	8/NOV/91	10	25
000 -010	SHIP STRUCTURAL STEEL	1/APR/91	10/JUN/91	3/MAY/91	12/JUL/91	25	50
010 -012	START ERECTING STRUCTURAL STEEL	15/JUL/91	5/AUG/91	26/JUL/91	16/AUG/91	10	15
012 -017	FINISH ERECTING STRUCTURAL STEEL	29/JUL/91	19/AUG/91	9/AUG/91	30/AUG/91	10	15
016 -019	START ROOF & DECKING	9/SEP/91	28/OCT/91	20/SEP/91	8/NOV/91	10	35
019 -021	FINISH ROOF & DECKING	23/SEP/91	11/NOV/91	18/OCT/91	6/DEC/91	20	0
018 -020	START INSTALLING CEILING	14/OCT/91	14/OCT/91	8/NOV/91	8/NOV/91	20	0
020 -021	FINISH INSTALLING CEILING	11/NOV/91	11/NOV/91	6/DEC/91	6/DEC/91	20	0
001 -005	START FLOOR ELECTRICAL	22/APR/91	22/APR/91	10/MAY/91	10/MAY/91	15	0
005 -007	FINISH FLOOR ELECTRICAL	13/MAY/91	13/MAY/91	14/JUN/91	14/JUN/91	25	0
010 -011	START WALL ELECTRICAL	15/JUL/91	7/OCT/91	9/AUG/91	1/NOV/91	20	60
011 -015	FINISH WALL ELECTRICAL	12/AUG/91	4/NOV/91	30/AUG/91	22/NOV/91	15	60
015 -021	INSTALL CEILING ELECTRICAL	2/SEP/91	25/NOV/91	13/SEP/91	6/DEC/91	10	60
001 -004	START INSTALLING PLUMBING	22/APR/91	13/MAY/91	10/MAY/91	31/MAY/91	15	15
004 -007	FINISH INSTALLING PLUMBING	13/MAY/91	3/JUN/91	24/MAY/91	14/JUN/91	15	15
003 -010	BUILD WALL FRAMES	6/MAY/91	17/JUN/91	31/MAY/91	12/JUL/91	20	20
010 -013	START FRAMEWORK	15/JUL/91	15/JUL/91	9/AUG/91	9/AUG/91	20	0
013 -017	FINISH FRAMEWORK	12/AUG/91	12/AUG/91	30/AUG/91	30/AUG/91	15	0
017 -018	FINISH WOODWORK	2/SEP/91	2/SEP/91	11/OCT/91	11/OCT/91	30	0
018 -021	START PAINT	14/OCT/91	25/NOV/91	25/OCT/91	6/DEC/91	10	30
021 -022	FINISH PAINT	9/DEC/91	9/DEC/91	3/JAN/92	3/JAN/92	20	0
022 -023	FINAL APPROVAL	6/JAN/92	6/JAN/92	10/JAN/92	10/JAN/92	5	0
021 -023	CLEAN SITE	9/DEC/91	16/DEC/91	3/JAN/92	10/JAN/92	20	5

Figure 6-8   Project schedule.

required may create a problem to hold the project duration within 41 weeks) we will identify the causes of the problem and offer alternatives to contain the planned project duration at 41 weeks in the event of excessive resource demands.

Each activity has an associated time estimate. To complete these activities within the specified time, certain resources are made available, which are specified and noted for each activity as shown in the planning diagram (Figure 6-6). For instance, it is indicated that it will take 5 weeks to complete "Finish Floor Electrical." This is an estimate based on one laborer and three electricians doing the job. Note how the critical trades required to do each activity are shown on the diagram. (In this example, only those resources that are considered critical are shown on the diagram.) Resources other than skilled and unskilled labor, such as critical facilities and equipment, needed to complete the job should also be shown on the diagram.

The next step is to input the resource data into the software program. The completed input data, consisting of the critical resources, are reproduced on the screen (see Figure 6-9). As noted on the screen, the critical resources identified for further investigation regarding allocation are: Carpenter, Electrician, Laborer, Ironworker, and Equipment Operator. The critical-trades availability can also be reported as a summary basis. Figure 6-10 shows a partial printout of a daily summary of the supply of the critical trades. Most software can also provide summaries of the supply of critical trades on a weekly and monthly basis.

Another computer printout (Figure 6-11) that will be helpful in the eventual resource-leveling analysis is the tabulation that combines the project activity (early start) schedule with the skilled-trades demands of each activity.

```
                                   Plan:BLDGPROJ.20  Proj:BLDGPROJ    MENU
Change  Task  Window  Screen  Files  Plan  Options  Default  Quit
Erase, Resource
┌════════════════════════════════Resource Report═══════════════════════ R═┐
│ Res▸ ──────────Description─────────▸  Calendar▸  Loading▸ Type▸ Supply─▸  │
│  CA   CARPENTER                       DAILY      Normal  Unlmtd  3        │
│  CT   COST $(00)                      DAILY      Normal  Limtd   Infinite │
│  EL   ELECTRICIAN                     DAILY      Normal  Unlmtd  3        │
│  EP   EQUIPMENT OPERATOR              DAILY      Normal  Unlmtd  2        │
│  LA   LABORER                         DAILY      Normal  Unlmtd  7        │
│  MW   IRONWORKER                      DAILY      Normal  Unlmtd  4        │
│                                                                          │
│                                                                          │
│                                                                          │
│  ┌──────────────────────────┬────────────────────────────────────────┐  │
│  │ Start Date▸  Supply─▸     │ Res▸ ──Qty──▸                          │  │
│  │                          │                                        │  │
│  │                          │                                        │  │
│  │                          │                                        │  │
└──┴──────────────────────────┴────────────────────────────────────────┴──┘
```

                                                CALC OVR NUM CAPS SCROLL

Figure 6-9 Identification of available (critical) resources for Building Project.

REPORT TYPE   : SUPPLY SUMMARY-CRITICAL TRADES - Supply
TIME NOW DATE: 1/APR/91   PRODUCED BY: P1211

	CA	CT	EL	EP	LA	MW
1/APR/91	3.00	·	3.00	2.00	7.00	4.00
2/APR/91	3.00	·	3.00	2.00	7.00	4.00
3/APR/91	3.00	·	3.00	2.00	7.00	4.00
4/APR/91	3.00	·	3.00	2.00	7.00	4.00
5/APR/91	3.00	·	3.00	2.00	7.00	4.00.
6/APR/91	*	*	*	*	*	*
7/APR/91	*	*	*	*	*	*
8/APR/91	3.00	·	3.00	2.00	7.00	4.00
9/APR/91	3.00	·	3.00	2.00	7.00	4.00
10/APR/91	3.00	·	3.00	2.00	7.00	4.00
11/APR/91	3.00	·	3.00	2.00	7.00	4.00
12/APR/91	3.00	·	3.00	2.00	7.00	4.00
3/JAN/92	3.00	·	3.00	2.00	7.00	4.00
4/JAN/92	*	*	*	*	*	*
5/JAN/92	*	*	*	*	*	*
6/JAN/92	3.00	·	3.00	2.00	7.00	4.00
7/JAN/92	3.00	·	3.00	2.00	7.00	4.00
8/JAN/92	3.00	·	3.00	2.00	7.00	4.00
9/JAN/92	3.00	·	3.00	2.00	7.00	4.00
10/JAN/92	3.00	·	3.00	2.00	7.00	4.00
11/JAN/92	*	*	*	*	*	*

Figure 6-10   Summary of daily supply of critical trades.

RESOURCE-LEVELING SCHEDULING

The initial leveling exercise is to assess the effect of the demand of the critical resources when using the early start schedule. The computer program schedules the activities within the framework of the schedule and attempts to initialize the available supply of critical resources so that what is available will not be exceeded. Leveling with the computer calculation approach is essentially the same as the manual method—schedule adjustments to the activities with float times.

Since most of the jobs require more than one resource, the decision as to which jobs to schedule early start and which jobs to defer becomes an exercise in the leveling procedure. For example, when a job is deferred, it potentially defers every other job in the project that follows it in a logical sequence. Without computer assistance, leveling can be a long and tedious task.

BUILDING PROJECT

REPORT TYPE : ACTIVITY SCHEDULE&REQUIREMENTS : Normal
TIME NOW DATE: 1/APR/91   PRODUCED BY: P1211

Node Numbers	Description	CA.	LA.	MW	EL	EP	Early Start	Early Finish
000 -001	EXCAVATE	0	90	0	0	30	1/APR/91	19/APR/91
000 -002	PREPARE REINFORCING STEEL	0	0	40	0	0	1/APR/91	12/APR/91
002 -006	PLACE REINFORCING STEEL	0	10	40	0	10	22/APR/91	3/MAY/91
006 -007	TIE REINFORCING STEEL	0	20	40	0	10	6/MAY/91	17/MAY/91
000 -003	BUILD FLOOR FORMS	30	30	0	0	0	1/APR/91	12/APR/91
003 -007	ERECT FORMS	30	45	0	0	15	22/APR/91	10/MAY/91
007 -009	CHECK FORMS	5	0	5	5	0	17/JUN/91	21/JUN/91
007 -008	PLACE FLOOR CONCRETE	20	40	0	0	0	17/JUN/91	28/JUN/91
017 -020	CHECK WALLS	5	0	0	5	0	2/SEP/91	6/SEP/91
008 -010	REMOVE FORMS	20	40	0	0	0	1/JUL/91	12/JUL/91
009 -010	CHECK FLOOR	5	0	5	5	0	1/JUL/91	5/JUL/91
010 -014	START MASONRY	0	80	0	0	0	15/JUL/91	9/AUG/91
014 -016	FINISH MASONRY	0	40	0	0	0	12/AUG/91	6/SEP/91
016 -020	CEMENT FINISH	0	20	0	0	0	9/SEP/91	20/SEP/91
000 -010	SHIP STRUCTURAL STEEL	0	0	0	0	0	1/APR/91	3/MAY/91
010 -012	START ERECTING STRUCTURAL STEE	0	0	40	0	10	15/JUL/91	26/JUL/91
012 -017	FINISH ERECTING STRUCTURAL STE	0	0	40	0	10	29/JUL/91	9/AUG/91
016 -019	START ROOF & DECKING	0	40	30	0	10	9/SEP/91	20/SEP/91
019 -021	FINISH ROOF & DECKING	0	80	60	0	40	23/SEP/91	18/OCT/91
018 -020	START INSTALLING CEILING	40	80	0	0	20	14/OCT/91	8/NOV/91
020 -021	FINISH INSTALLING CEILING	40	80	0	0	20	11/NOV/91	6/DEC/91
001 -005	START FLOOR ELECTRICAL	0	15	0	0	0	22/APR/91	10/MAY/91
005 -007	FINISH FLOOR ELECTRICAL	0	25	0	75	0	13/MAY/91	14/JUN/91
010 -011	START WALL ELECTRICAL	0	20	0	60	0	15/JUL/91	9/AUG/91
011 -015	FINISH WALL ELECTRICAL	0	15	0	30	0	12/AUG/91	30/AUG/91
015 -021	INSTALL CEILING ELECTRICAL	0	10	0	20	0	2/SEP/91	13/SEP/91
001 -004	START INSTALLING PLUMBING	0	15	0	0	0	22/APR/91	10/MAY/91
004 -007	FINISH INSTALLING PLUMBING	0	20	0	0	0	13/MAY/91	24/MAY/91
003 -010	BUILD WALL FRAMES	20	80	0	0	20	22/APR/91	17/MAY/91
010 -013	START FRAMEWORK	60	40	0	0	20	15/JUL/91	9/AUG/91
013 -017	FINISH FRAMEWORK	30	60	0	0	15	12/AUG/91	30/AUG/91
017 -018	FINISH WOODWORK	120	60	0	0	0	2/SEP/91	11/OCT/91
018 -021	START PAINT	0	10	0	0	0	14/OCT/91	25/OCT/91
021 -022	FINISH PAINT	0	20	0	0	0	9/DEC/91	3/JAN/92
022 -023	FINAL APPROVAL	5	0	0	5	0	6/JAN/92	10/JAN/92
021 -023	CLEAN SITE	0	80	0	0	20	9/DEC/91	3/JAN/92
TOTAL		430	1165	300	205	250		

Figure 6-11 Project schedule (early start) with resource allocation (labor-weeks) per activity.

Following the same approach for leveling as the manual method, the computer is usually programmed (defaulted) for this sequence: assign the resources for those jobs on the critical path; assign resources for those jobs with minimal float time; and schedule the remaining jobs with maximum float times.

In every case we have to assume that there are enough personnel available for work on the critical jobs, so that critical jobs will be scheduled first. Noncritical jobs will then be scheduled, with the resources that remain. If there are so many noncritical jobs that cannot be completed on time with the limited sources available, a decision will be necessary as to whether the project duration will need to be extended beyond the time predicated for completion by the original timing solution. Other alternatives are overtime, or possibly a revision of the timing plan, extending the project date. The leveling procedure compares the resource requirements with the available resources and decides how best to distribute the resources to accomplish the project in the time required.

There may be projects where the available resources are not on a constant level throughout the project, and the computer programs can usually accommodate changes in availability of labor. Therefore, it will be necessary to specify the resource demands on a period-by-period basis, allowing the levels of availability to be changed. This does permit some flexibility, since it is quite usual that several projects go on at the same time, and by specifying the resource levels on a period-by-period basis, it is often possible to transfer resources from one project to another, which will result in efficient projects. One of the major benefits is that this will also tend to reduce project expenditures.

For this sample problem, a resource summary was requested after we completed the leveling process. The daily resource demand summary is shown in part in Figure 6-12. (Only the five critical trades are shown in this tabulation

BUILDING PROJECT

REPORT TYPE   : DEMAND SUMMARY-CRITICAL TRADES - Demand
TIME NOW DATE: 1/APR/91   PRODUCED BY: P1211

	CA	CT	EL	EP	LA	MW
1/APR/91	-	66	-	2.00	6.00	4.00
2/APR/91	-	66	-	2.00	6.00	4.00
3/APR/91	-	66	-	2.00	6.00	4.00
4/APR/91	-	66	-	2.00	6.00	4.00
5/APR/91	-	66	-	2.00	6.00	4.00
6/APR/91	*	*	*	*	*	*
7/APR/91	*	*	*	*	*	*
8/APR/91	-	66	-	2.00	6.00	4.00
9/APR/91	-	66	-	2.00	6.00	4.00
10/APR/91	-	66	-	2.00	6.00	4.00
11/APR/91	-	66	-	2.00	6.00	4.00
12/APR/91	-	66	-	2.00	6.00	4.00
13/APR/91	*	*	*	*	*	*
14/APR/91	*	*	*	*	*	*
15/APR/91	-	41	-	2.00	6.00	-
16/APR/91	-	41	-	2.00	6.00	-
17/APR/91	-	41	-	2.00	6.00	-
3/JAN/92	-	6	-	1.00	5.00	-
4/JAN/92	*	*	*	*	*	*
5/JAN/92	*	*	*	*	*	*
6/JAN/92	1.00	10	1.00	-	-	-
7/JAN/92	1.00	10	1.00	-	-	-
8/JAN/92	1.00	10	1.00	-	-	-
9/JAN/92	1.00	10	1.00	-	-	-
10/JAN/92	1.00	10	1.00	-	-	-
11/JAN/92	*	*	*	*	*	*

Figure 6-12 Summary of daily critical-trades requirements.

chart.) The time periods are listed down the page so that one can readily identify the resources needed at any particular time period. (For convenience, to show both the daily summary of resource requirements and resource availability, the computer printouts were reproduced, with some modification.)

Most computer programs can provide a printout showing a graphical presentation (or source chart) of resource requirements by plotting the type of resource against time. The computer program can generate histograms that will display both resource availability and demand.

When using the earliest start schedule, the resource charts show that demand exceeds the supply of carpenters, laborers, and equipment operators at various times during the course of the project. This schedule also reflects demand variations ("peaks and valleys") during the course of the project that are not desirable.

However, there may be important reasons why the duration of this project does not change. Therefore, the leveling exercises for this project will be confined within the original project duration. Most computer software also permits extending the project duration to satisfy the resource supply if actual project conditions permit.

The leveling exercise that is used follows the approach discussed previously, that is, scheduling the resources based on float times where available resources are allocated first to the critical path items. The particular software

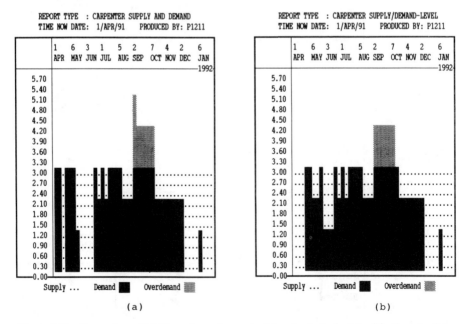

Figure 6-13 Carpenter availability and requirements: (a) early start schedule; (b) schedule after leveling.

program that we are using for this type of exercise will reschedule the activities within the stated supply of resources.

As we have already stated, for this sample problem we have established that the schedule of some of the jobs must change. The way this is most commonly done is to (1) do the jobs on the critical path; and (2) with any resources left over, do the jobs with the smallest total float time. The computer program is basically a trial-and-error process and the leveling is a function of the order of the activity's float times in the input.

Figures 6-13 through 6-17 compare the critical-trades requirements when the Building Project is scheduled first, with the early start, and then, after the leveling process. Although this leveling procedure improves the resource supply and demand picture only to a certain extent, it is a positive contribution. However, this first leveling attempt, which contains the original project duration, does not allow for leveling in its truest sense, as there are peaks and valleys and at times, demand exceeds supply on several of the critical trades. A "snapshot" summary of demand versus supply is shown in the following table (more detail is provided in Figures 6-13 through 6-17).

| | | Demand More Than Supply | | | |
| | | Early Start Schedule | | After Leveling Schedule | |
	Supply	Number	Weeks	Number	Weeks
Carpenter	3	2	1	1	6
		1	4		
Laborer	7	5	1	3	1
		4	2	1	4
		2	5		
		1	1		
Equipment Operator	1	1	3	1	3

Leveling allows for a better distribution of the skilled labor force. (Note the histograms on better distribution of laborers and potential overtime reduction after project leveling.) However, leveling will not reduce the total amount of hours required for each trade to perform the project tasks.

It is important to understand that the majority of construction projects require specific trade classifications. In other words, one person cannot perform more than one trade (a by-product of the skilled trades union efforts in the United States). Therefore, no worker can be assigned to do *any* type of work. Consequently, resource requirements need to be considered by individual trades or classifications; plotting total numbers of personnel has very little significance. Handling the individual crafts separately is required, because in most cases, workers cannot be moved freely from one craft classification to another.

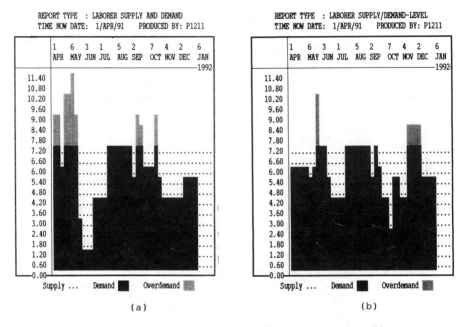

Figure 6-14 Laborer availability and requirements: (a) early start schedule; (b) schedule after leveling.

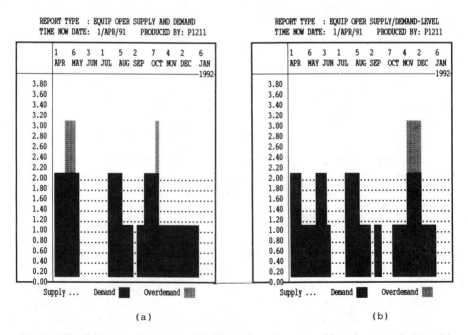

Figure 6-15 Equipment operator availability and requirements: (a) early start schedule; (b) schedule after leveling.

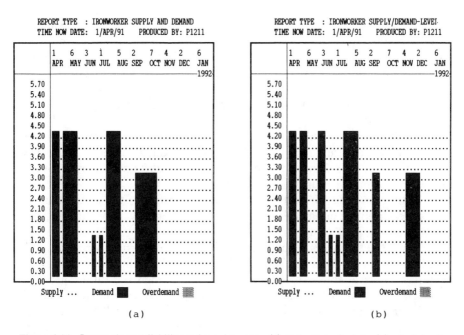

Figure 6-16 Ironworker availability and requirements: (a) early start schedule; (b) schedule after leveling.

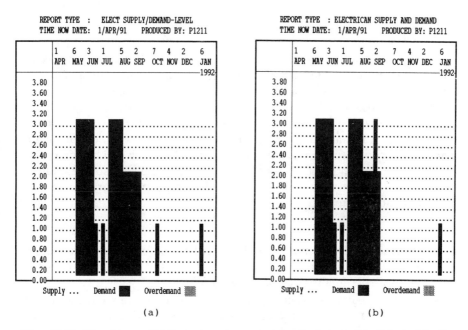

Figure 6-17 Electrician availability and requirements: (a) early start schedule; (b) schedule after leveling.

In every case we have to assume that there are enough personnel available for work on the critical jobs and period by period; critical jobs will therefore be scheduled first. Noncritical jobs will then be scheduled with the resources that remain. However, there may be so many noncritical jobs that cannot be completed on time with the limited resources available that the project duration would be extended beyond the time predicted for completion by the original timing solution.

For this sample problem, the planned project completion date is expected to be contained, so adjustments will be made to the schedule of individual activities. Figure 6-18 shows the after-leveling schedule for each activity for this sample problem by holding the completion date.

If after leveling within the project time constraint, the number of resources required are intolerably greater than the supply available, the decision to lengthen the project duration becomes real.

When the resources required are unavailable at the time a critical job is scheduled to start, the start is delayed to the earliest time when adequate resources are available. This, of course, will lengthen the duration of the project. Depending on the solution, some of the jobs that were not originally on the critical path may have become critical. A new schedule will be required to take into account the extended project duration date. The extended schedule, in the form of a bar chart generated by computer software, has been worked up to show the timing differences from the original schedule.

SUMMARY

A complete planning process must include time and cost factors, which have to relate satisfactorily to the project schedule. An additional factor must be considered: the resources available to do the jobs, such as labor, equipment, space, and funds.

A resource-leveling program, whether done manually or by using a computer program, allocates the resources available for a project in such a manner that period-to-period changes in the resource levels will be minimized and resource demand will approach resource supply. The resource-leveling program can also be used to determine a minimum project duration given a limited quantity of resources.

The leveling of resources involves these actions:

- The resource levels: the number of construction or tradespeople, engineers, and programmers; the number of types of equipment required; and available funds specified for each job.
- In the computer program, a printout will show the skills and equipment required based on the project schedule after leveling.
- With the availability of each resource given period by period, the computer then attempts to schedule the jobs and equipment within the

REPORT TYPE : ACTIVITY SCHEDULE&REQUIREMENTS : Normal
TIME NOW DATE: 1/APR/91   PRODUCED BY: P1211

Node Numbers	Description	CA.	LA.	MW	EL	EP	Early Start	Early Finish
000 -001	EXCAVATE	0	90	0	0	30	1/APR/91	19/APR/91
000 -002	PREPARE REINFORCING STEEL	0	0	40	0	0	1/APR/91	12/APR/91
002 -006	PLACE REINFORCING STEEL	0	10	40	0	10	22/APR/91	3/MAY/91
006 -007	TIE REINFORCING STEEL	0	0	0	0	0	6/MAY/91	17/MAY/91
000 -003	BUILD FLOOR FORMS	0	0	0	0	0	1/APR/91	12/APR/91
003 -007	ERECT FORMS	30	45	0	0	15	6/MAY/91	24/MAY/91
007 -009	CHECK FORMS	5	0	5	5	0	17/JUN/91	21/JUN/91
007 -008	PLACE FLOOR CONCRETE	20	40	0	0	0	17/JUN/91	28/JUN/91
017 -020	CHECK WALLS	0	0	0	0	0	2/SEP/91	6/SEP/91
008 -010	REMOVE FORMS	20	40	0	0	0	1/JUL/91	12/JUL/91
009 -010	CHECK FLOOR	5	0	5	5	0	1/JUL/91	5/JUL/91
010 -014	START MASONRY	0	80	0	0	0	15/JUL/91	9/AUG/91
014 -016	FINISH MASONRY	0	40	0	0	0	12/AUG/91	6/SEP/91
016 -020	CEMENT FINISH	0	0	0	0	0	9/SEP/91	20/SEP/91
000 -010	SHIP STRUCTURAL STEEL	0	0	0	0	0	1/APR/91	3/MAY/91
010 -012	START ERECTING STRUCTURAL STEE	0	0	40	0	10	15/JUL/91	26/JUL/91
012 -017	FINISH ERECTING STRUCTURAL STE	0	0	40	0	10	29/JUL/91	9/AUG/91
016 -019	START ROOF & DECKING	0	40	30	0	10	9/SEP/91	20/SEP/91
019 -021	FINISH ROOF & DECKING	0	0	0	0	0	23/SEP/91	18/OCT/91
018 -020	START INSTALLING CEILING	40	80	0	0	20	14/OCT/91	8/NOV/91
020 -021	FINISH INSTALLING CEILING	40	80	0	0	20	11/NOV/91	6/DEC/91
001 -005	START FLOOR ELECTRICAL	0	15	0	0	0	22/APR/91	10/MAY/91
005 -007	FINISH FLOOR ELECTRICAL	0	25	0	75	0	13/MAY/91	14/JUN/91
010 -011	START WALL ELECTRICAL	0	20	0	60	0	15/JUL/91	9/AUG/91
011 -015	FINISH WALL ELECTRICAL	0	15	0	30	0	12/AUG/91	30/AUG/91
015 -021	INSTALL CEILING ELECTRICAL	0	10	0	20	0	2/SEP/91	13/SEP/91
001 -004	START INSTALLING PLUMBING	0	15	0	0	0	22/APR/91	10/MAY/91
004 -007	FINISH INSTALLING PLUMBING	0	20	0	0	0	13/MAY/91	24/MAY/91
003 -010	BUILD WALL FRAMES	10	40	0	0	10	6/MAY/91	31/MAY/91
010 -013	START FRAMEWORK	60	40	0	0	20	15/JUL/91	9/AUG/91
013 -017	FINISH FRAMEWORK	30	60	0	0	15	12/AUG/91	30/AUG/91
017 -018	FINISH WOODWORK	120	60	0	0	0	2/SEP/91	11/OCT/91
018 -021	START PAINT	0	10	0	0	0	14/OCT/91	25/OCT/91
021 -022	FINISH PAINT	0	20	0	0	0	9/DEC/91	3/JAN/92
022 -023	FINAL APPROVAL	5	0	0	5	0	6/JAN/92	10/JAN/92
021 -023	CLEAN SITE	0	80	0	0	20	9/DEC/91	3/JAN/92
TOTAL		385	975	200	200	190		

Figure 6-18 Project schedule (after leveling) with resource allocation (labor-weeks) per activity.

framework of the planning network so that specified availability is not exceeded. If this is impossible, the program will extend the length of time for the project until a feasible schedule is obtained.

The resource-leveling program is a valuable tool for the project planner. The program can provide many answers to resource-leveling problems; however, the solution based on resource leveling needs to be compared with other factors, such as cost, timing, and other management objectives, before arriving at a desirable plan.

7
The Role of the Computer

Project management is basically a method of planning and scheduling projects through the use of various tools that are presently available. One tool that has become an absolutely necessary expedient in implementing project management techniques is the computer. By handling routine calculations faster and with uncanny accuracy, the computer can help plan and analyze projects faster than manual methods. *However, it cannot plan or analyze the project* without the logical ability of the human being.

As noted in previous chapters, a manual approach to project planning and scheduling of time, cost, and personnel is possible, but it is laborious and tedious even for very small projects. Once the computer approach is understood, its application becomes more desirable for all projects, especially those that are complex and relatively large. In general, when a question arises as to whether a computer is required to apply project management techniques to a project, the decision is made not only on the size and complexity of the project, but also on the type of analysis required and the frequency of the updating requirements. Use of the computer will be most favored.

Trade-offs need to be considered regarding the time required to perform calculations manually versus the costs associated with the procurement of a computer system and of training personnel and launching the system. However, when it becomes apparent that computer calculations for larger projects may take minutes whereas manual calculations for those projects may require days, there is no contest. In addition, the features of the computer software will provide a great deal more useful information for analysis purposes. Complexity is another factor to be considered, as performing manual calculations will not fulfill the requirements of a complex project. Simple timing calculations derived manually may not provide the analysis for time/cost trade-offs when project timing is an issue or for resource planning when leveling skilled trades or professional personnel becomes an issue and "what-if" situations are necessary.

The frequency with which the project status is to be updated may also determine the need for using computer-generated information. Frequent manual updating to process data quickly and accurately will undoubtedly occupy more time than may be practicable.

Before discussing how a computer is used in planning, scheduling, and controlling projects, we will briefly review both the physical elements of the computer, known as *hardware,* and *software,* the instructions given the computer.

COMPUTER HARDWARE

Computers vary tremendously in size and capability; however, they all operate similarly as an integrated system. All of the units are under the control of the central processor and will operate as noted in the following layout:

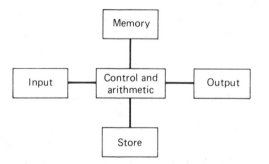

Input devices feed the computer with data. These devices include visual display terminals, links with other computers, console keyboards, magnetic tape, and cassettes. Input devices are also known as *peripherals.* Peripherals are devices that enhance a basic computer configuration, providing the system with varying degrees of sophistication. In addition to input, peripherals are associated with output and auxiliary storage devices, including magnetic tape and magnetic disks.

Storage and memory (data base) comprise the file store of the computer, making vast amounts of information available to the processor in an unbelievably short time, measured in millionths of a second. The computer can look up information held in storage and can add to or erase information that is stored. On a computer drive where the data are filed, you can access information wherever it appears, anywhere on the surface of the disk. This is known as random access, and the time needed for retrieval is becoming shorter and shorter.

The *central processor* performs all calculations, simple logic tasks, and all input and output to the system. As the heart of the computer, it draws instructions one by one from the memory and puts the corresponding machine operations into effect. By connecting all the units of the system, it directs the transfer of information. The memory unit has a twofold purpose: to act as a temporary store for data and to hold the program that directs the operations of the computer.

Output is where the results come out. Frequently, this is a printing device, but other units are available, including visual display. In many systems, output is to a terminal computer that prints out the results of inquiries made of the central system. Printed output units, including carriage, serial, and line printers, are also known as peripherals.

COMPUTER SOFTWARE (TELLING THE COMPUTER WHAT TO DO)

Before any computer can start work, it must be provided with detailed instructions. The process of translating a program into instructions that the computer can understand and use is called *coding* or *programming*. These programs constitute the *software* of the system. Computers need instruction to perform even the simplest task. A set of instructions is called a *program*. Before preparing a program, the programmer makes a *flowchart* to show the sequence of operations.

Two types of languages are used in programming: machine language and coded language. A *machine language* consists of numerical symbols that are unique to a given type of computer. This is the only language that a computer understands. Programs written in machine language are very detailed and extensive, requiring many hours of tedious work to write.

OPERATING SYSTEMS

Operating systems link hardware and applications software to a specific microprocessor. They are critical to the operation of any computer because they determine which programming language and application programs can be used. The operating system package typically contains a number of utility programs for formatting disks, copying data or program files from one disk to another, displaying a list of programs or data on a disk, editing programs and data, and checking the contents of disks for faults. Several popular packages are listed here.

Apple DOS is a proprietary disk operating system written for the 8-bit Apple computer with a Mostek 6502 microprocessor. Apple DOS, TRS-DOS, and MS-DOS are not compatible.

CP/M (Control Program for Microprocessors), developed by Digital Research Corp., Pacific Grove, California, for Zilog Z80 and Intel 8080 microprocessors, has become the standard operating system for 8-bit microcomputers. More applications programs are said to have been written to run on it than on any other operating system.

MS-DOS is distributed by Microsoft Corp., Bellvue, Washington, for use with Intel 8088 microprocessors. Because of the huge amount of software written for it, this operating system is considered by some to be the standard for 16-bit microcomputers. Its popularity is associated with PC-DOS, a version of MS-DOS for the IBM PC microcomputer.

PC-DOS is a version of MS-DOS for the IBM PC microcomputer.

TRS-DOS is a proprietary disk operating system for Tandy/Radio Shack 8-bit microcomputers. TRS-DOS, Apple DOS, and MS-DOS are not compatible.

UNIX, developed by Bell Laboratories in 1960 for use on minicomputers such as the DEC PDP-11, allows certain application programs written on minicomputers to be transferred to the larger microcomputers, with minimal changes. It runs on Intel 8086, Motorola MC68000, and Zilog Z8000 microprocessors and allows more than one user to run several programs simultaneously. Several versions are available.

UCSD p-System, developed at the University of California at San Diego, fosters software portability. Compilers for the p-System produce codes for a pseudo or p-machine. Then brief programs for each type of microcomputer translate the p-code into the machine code required by a specific microcomputer.

PROGRAMMING LANGUAGES

A programming language is a translator program that converts the source code of an application program into object code or machine language. The most useful programming languages for plant engineering are BASIC, FORTRAN, and Pascal.

APL (A Programming Language) is a general-purpose language that is self-teaching and interactive. It is a very powerful language, requiring fewer instructions for complex operations than do most other languages. Beginners are said to be able to produce simple programs after only a few hours' work with the language. APL is an interpreted language.

BASIC (Beginners All-Purpose Symbolic Instruction Code) is the most popular language for microcomputer programming. It is a fairly easy language for nonprogrammers to learn and remember because many of the commands are English words. It does not offer the power of the more engineering-oriented languages, but it can be used for a variety of engineering applications. Although BASIC can be compiled, it is normally used as an interpreted language.

COBOL (COmmon Business-Oriented Language) uses natural words and phrases that can be recognized by nontechnical users. It offers limited mathematic capabilities but works well for manipulating text and data. COBOL is a compiled language.

FORTRAN (FORmula TRANslator) was one of the first high-level programming languages. Although not as efficient as newer languages, it is one of the more widely used languages for engineering work because such a large number of engineering programs have been written in it. FORTRAN is large-computer oriented, but versions are available for the larger microcomputers. It is a compiled language.

Pascal was designed to speed the writing of large, complex programs. It is a highly readable language that encourages structured programming. It is often used in education and research. Pascal is a compiled language.

COMPUTER APPLICATION

As we have discussed previously, the computer does not necessarily have to be used (but it is becoming increasingly feasible to use a computer). Included among the factors that determine when a computer should be considered for use are the following:

- Size of the project
- Complexity of the project
- Type of analysis required
- Frequency with which the project needs to be updated

Size of the project.   A computer can do in minutes what may take days using manual calculations. Remember that the computer needs data input, and sometimes the nature of the input, with its corrections, can take a great deal of time, which could offset the time savings over manual methods.

Complexity.   A complex program may be impossible or at least very difficult to calculate manually. (However, time-consuming input data preparation can overshadow the benefits derived from computer calculations. You can get bogged down with computer data sheets, "bugs," and other factors.)

Type of analysis.   Resource leveling and time/cost trade-offs, which contain a number of variables, can be done much more readily with a computer.

When you need different sorts or categories, by responsibility, float, or early start, a computer wins hands down.

Frequency. Once you get the hang of preparing computer input as described for updating, you will normally find using a computer more desirable than making manual updates.

The value of the computer as a management tool in calculating schedules is evident when considering the factors listed above. Using the computer for project control offers the following specific benefits:

- Involves more project teams in up-front planning and project control.
- Better project schedule control (which saves cost by reducing overtime and minimizing extended project durations).
- Increased communication through easy-to-understand reports, and increased communication among departments or organizations involved with a project.
- Regular updating requires minimum personnel requirements.
- The time to produce updates and status reports is reduced substantially.
- Much manual effort, which is vulnerable to error, is eliminated.
- Updating programs means that added work activities may have to be inserted and others deleted. Using a computer minimizes the effort needed for this type of work at any stage during the course of a project.
- Risks associated with project delays are reduced. "What-if" analyses can be done on a timely basis well ahead of potential problems (an early-warning system).
- And the project teams have a disciplined method to use to think through the project.

COMPUTER PROGRAMS

There are well over 200 project management program packages available for use with personal computers, offered by firms dealing with all types of software and by firms that specialize in project management software. There are also project management software packages available that are prepared by computer hardware firms for users of their computer equipment. Competition has mandated that all of these programs be completely tested, ready for use, and updated and improved periodically. The contents of a software package should include a program description and an instruction manual. Time and effort on the part of the user will still be necessary to acquire a working knowledge of the package. Although the majority of programs are "user friendly," it may be wise to invest in a training session conducted by someone who has used the program extensively, possibly by a qualified person from the software firm.

The following representative list of project management software presently on the market is only for information and guidance. No specific recommendation is being made. The specific needs of the project under consideration should determine the choice. There is no single best package; major considerations include easy-to-read documentation, ease of use (sometimes called user friendliness), and flexibility in preparing reports, such as support of both arrow and precedence diagramming. Special features supporting resource allocation, such as handling "what-if" exercises and cost analysis, should also be considered in the selection.

Personal Computer Software	Product	Approximate List Price
APPLIED BUSINESS TECHNOLOGY 361 Broadway New York, NY 10013	Project Workbench Advanced	$1,250
COMPUTER ASSOCIATES INTERNATIONAL, INC. 12540 McKay Drive San Jose, CA 95131	Super Project Plus	495
COMPUTERLINE, INC. 52 School Street Pembroke, MA 02330	Plan TRAC II	1,000
METLER MANAGEMENT SYSTEMS 435 Devon Park Drive Suite 300 Wayne, PA 19087	Prestige	2,995
METIER MANAGEMENT SYSTEMS, INC. 2500 N. Look Wish Suite 1300 Houston, TX 77022	Artemis Project	3,500
MICROSOFT CORP. 1011 NE 35th Way Richmond, WA 94073	Microsoft Project	495
PRIMEVERA SYSTEMS, INC. 2 Bala Avenue Bala Cyniaya, PA 19004	Primevera Project Planner	2,500
PROJECTRONICS, INC. 4546 El Camino Real Suite 324 Los Altos, CA 94022	Pertmaster Advance	1,495
PROJECT SOFTWARE & DEVELOPMENT, INC. 20 University Road Cambridge, MA 02138	Quicknet Professional	1,575
SYMANICA 505 B. Saw Mariw Drive Novato, CA 94945	Timeline 3.01	595

All of the above run under the MS-DOS environment.

Mainframe, Minicomputer Software	Product	Approximate Price
METIER MANAGEMENT SYSTEMS, INC. 2500 N. Look Wish Suite 1300 Houston, TX 77022	Antemis 7000	$27,500
PROJECT SOFTWARE & DEVELOPMENT, INC. 20 University Road Cambridge, MA 02138	Project/2	1,675
SYSTONETICS, INC. 1561 E. Orangethrope Suite 200 Fullerton, CA 92631	Expert Vision	45,000
TECHNICAL ECONOMICS, INC. 1650 Solano Avenue Albany, CA 94707	Vire	30,000
UNISYS CORP. P.O. Box 500 Blue Bell, PA 19424	Optima 1100	30,000

There may be as many as 30 basic output reports that can be provided for basic information on the latest project status. (Many software packages have customizing features that can provide an innumerable number of various reports.) Reports are designed in various project formats suited for various levels of project managers and project analysts. There are also many types of cost reports for use by various financial groups.

A typical software package will produce many types of reports, including the following:

- *Scheduled dates:* A time schedule of all work items may be arranged in any number of sorts, such as earliest start, latest start, earliest finish, or latest finish dates. Schedules may also be sorted for total float time of each work item (and most programs can sort free and independent float times as well).

- *Bar chart:* Shows graphically the calculated duration of the specified work items, with total float times and the critical path also able to be noted on the bar chart.

- *Critical activities:* A sorting of critical activities along paths of zero and low float times.

- *Milestones:* A listing of major events in either start or finish dates. (Total float time can also be shown for each event.)

- *Responsibility (work breakdown structure):* A sorting of work items according to the participating group responsible for their completion.

- *Resource allocation:* A listing by each resource required for a given time span.
- *Resource leveling:* A graphical representation of the daily resource usage within the scheduled resource requirements.
- *Project costs:* A listing of the specified work items, together with the estimated total cost and the actual cost to date of each.
- *Cash flow schedules:* A listing of the specified work items, together with the estimated total cost and the actual cost to date for each, by calendar month.
- *Scheduled earnings:* A listing of the cost status for all specified work items, showing estimated cost, projected cost, and actual cost to date.
- *Cost optimization:* A time schedule of the work items that allows the project's duration to be shortened at minimum additional cost.
- *Work breakdown structure (WBS):* One of the most important features. If applied properly and if the software package contains this feature, it will allow the personal computer to approach the performance of several minicomputers and mainframe computers.

These reports represent the basic information that a typical software package of a project management and content system can provide. Reports can be designed to suit each level of management, project supervisors, project analysts, and financial analysts.

Software is continually changing, improvements being made to satisfy users as experience is gained in program use and as new programs are released. Costs for these programs vary, and details of purchase or lease should be discussed with the computer software firm.

In addition to standard reports, there are special software programs that produce project management graphical reports. Time-phased network diagrams can be plotted that display the flow of project activities and the overall impact of scheduling changes. Graphical reports can also include time-phased bar charts for reporting and monitoring schedule progress, as well as a wide variety of cost/resource graphs (cost, personnel requirements) used for budget and labor control.

Summarizing, personal computers are microcomputers, designed to be used by one person. This class of computers includes the hand-held, portable, transportable, and desktop models, and even supermicros, but is usually applied only to the transportable and desktop size. The term *personal computer* refers to a general-purpose desktop business machine capable of running popular business software (spreadsheets, word processors, data base managers, etc.), engineering calculation packages, and project management packages. A typical personal computer system configuration may contain the following:

	Basic System	Options
Hardware		
CPU	Personal computer (8088)	Math co-processor
Memory	265K-bytes RAM	512k-byte RAM
Storage	Two 360K-byte disk drives	Winchester hard disk
Monitor	Monochrome graphics monitor	Color graphics monitor
Interface	Monitor/Printer interface	Color graphics interface
Printer/Plotter	Dot matrix printer with graphics	Letter-quality printer pen plotter
Software		
System software	Standard operating system BASIC programming	Second operating system Complied language
General application software	Spreadsheet File/list manager Word processor	Integrated environmental software with graphics Data base manager
Engineering software	Equation solver	Calculation packages for thermodynamics, piping, stress analysis, etc.

A sample program, Computer Installation Project, will be used to illustrate the manner in which a computer is applied in the project management cycle of planning, scheduling, and controlling.

SAMPLE PROGRAM: COMPUTER INSTALLATION PROJECT

A flow chart depicting the steps in planning scheduling and controlling a project using project management techniques using the computer as a tool is shown in Figure 7-1.

Planning

The first step is to establish the objectives, which are usually influenced by company business plans and higher management directives. For this project it has been determined that the new computer system needs to be operational June 7, 1991, based on project approval (and start of project) on September 3, 1990.

After identifying the objectives, the next step is to plan what is to be done. The technique used is to prepare a work breakdown structure in conjunction with determining the listing of jobs necessary to complete the job.

Determining the Jobs Required

Proper listing of every step required to accomplish the project is of great importance. It is vital that all phases of work be encompassed by the jobs listed.

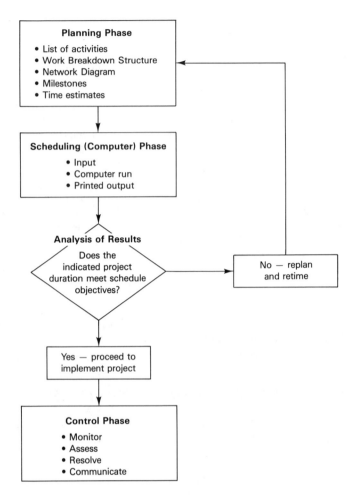

Figure 7-1   Flowchart for planning, scheduling, and controlling a project.

Any omissions will cause inaccuracies in scheduling and may result in failure to complete the project on time.

Figure 7-2 lists the detailed work involved in the computer project. Descriptions of the work should be phrased in such a way that they (1) state basic functions that can be understood by all concerned with managing the planning, scheduling, and control of the project; and (2) include all the work required in the project.

Once the activities list is completed, the next step is preparing the work breakdown structure (WBS). Figure 7-3 is a WBS diagram that shows the organizational structure among the activities involved in the computer project where they are categorized according to work interests (or departments).

With the initial objectives completed, together with the work activities list

The Associates Company has been installing computers for production control functions and this sample problem is a condensed version of the steps required to layout the program for the installation. The objective is to make the computer system operational by June 10, 1991, based upon start of project on September 3, 1990.

---

1. *Decide on computer:* Select computer configurations which will best fulfill needs.

2. *Procure computer:* Place order for selected computer, fabricating and delivery time.

3. *Install computer:* Physically install computer and related software.

4. *Determine site specifications:* Define physical environment for computer, including location, area size, environment requirements.

5. *Solicit bids for site preparation:* Prepare site layout and design, contact appropriate contractors for bids, provide major timing requirements.

6. *Award contract for site preparation:* Award contract based on specifications and proposal price.

7. *Prepare site:* Design and build computer enclosure in accordance with design specifications.

8. *Select programming personnel:* Interview and select personnel for computer operations based on experience, as with the selected computer and within project budget requirements.

9. *Train programming personnel:* Train for particular needs of project using the selected computer and applying system standards of the company.

10. *Select operating personnel:* Interview and select personnel for computer operations based on experience with the selected computer and within project budget requirements.

11. *Train operating personnel:* Train on computer system and on operating procedures of the company.

12. *Layout computer records:* Define and design files to fit the needs of the system.

13. *Develop computer program:* Design system appropriate for computer; code and test programs.

14. *Test computer program:* Test all programs run-to-run for continuity and balance back to manual controls.

15. *Design forms:* Define form needs and evaluate usage.

16. *Procure forms:* Order forms from appropriate outlet, giving volume requirements and needed date of receipt.

17. *Operate Program:* Complete computer and user documentation, train users, communication for ongoing maintenance.

Figure 7-2   List of activities for the Computer Installation Project.

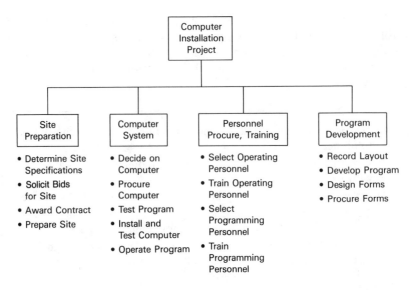

Figure 7-3   Work breakdown structure (WBS).

and the WBS, it may be appropriate to supplement the main objectives with any interim major events that need attention as the project proceeds. These interim events and the major objective are called *milestones*. Additional milestones for this project are as follows:

- Complete site preparation
- Computer on site

Figure 7-4 shows the completed network diagram that illustrates the interrelations among the activities in the project. Preparing the diagram (or planning model) involves developing a flow diagram depicting the logic used and the strategy behind the plan. An outline for preparing this model follows.

1. *Decide on computer* must be completed before:
   a. *Determine site specification* because these specifications would be determined in part by the type of computer purchased.
   b. *Select operating personnel* and *select programming personnel* because the qualifications required would be dependent to some extent on the type of computer.
2. *Test programs* can be started only *after* the completion of:
   a. *Install and test computer.*
   b. *Train operating personnel.*
   c. *Train programming personnel.*
   d. *Record layout.*
   e. *Develop program.*

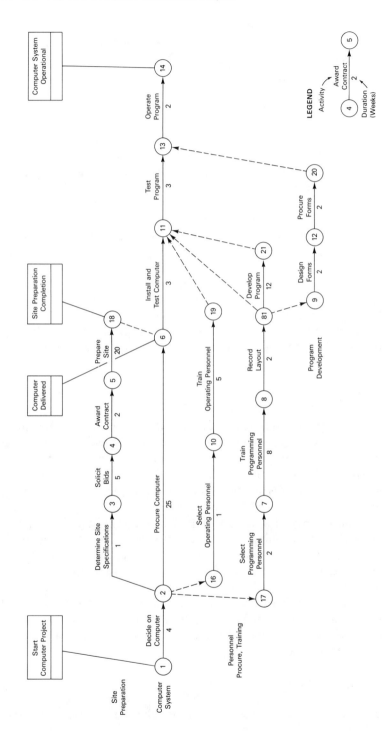

Figure 7-4 Planning diagram for the Computer Installation Project.

There is nothing unique about arrow diagramming—different people planning the same operation could arrive at a different plan of action, and therefore, different diagrams. For example, this computer project could have several different program development sequences, depending on the computer personnel who are planning the project. However, there are many sequences of jobs that are fairly rigid, and it would be difficult to replace them with any other satisfactory sequence.

Regardless of how a plan is set forth in an arrow diagram, it provides an effective graphic display of what it is planned to do, and allows for effective review. Time estimates for each job, made by knowledgeable persons, are shown beneath each job arrow. At this stage, planning the work sequence and estimating job times are done on a normal time basis, as the length of the project has not yet been determined. No effort is exerted at this point to compress project duration or to meet any particular deadline. If the output indicates that the project duration does not meet objectives, it will be necessary to replan the project and reduce some job times through expediting. At this point, simulation studies of alternative plans may require crash time/cost information for selected jobs to perform a time/cost compression analysis, and resource-leveling exercises may also be carried out.

After the "what has to be done" portion (planning phase) of the diagram is completed (and agreed to by all participants), computer scheduling can take place.

Scheduling

After the planning model has been agreed to by project participants, duration estimates for each activity can be set. (Setting durations can also be included in the planning phase, if desired.) It should be emphasized that the scheduling process is highly dependent on time estimates. If there is a weakness in any scheduling system, it is in time estimating. If people provide inaccurate time estimates, the total time will be inaccurate.

There is an orderly procedure to follow in analyzing these estimates. Briefly, this consists of examining jobs on the critical path first, then jobs on the next-longest path, and so on—questioning and revising time estimates that seem out of line until a satisfactory project schedule is developed.

For most projects, the relations between jobs or activities are not so obvious, particularly if it is necessary to go through every possible path and compute the timing of each job. The computations become very tedious and time consuming if the manual approach is used. Thus for most projects the use of a computer becomes a necessity.

Preparing Computer Input Data

The required input data are entered in a disciplined manner using interactive query languages (questions and answers), entry screen menus, or systems prompts (or there may be a combination of these). Interactive input features on-line, conversational means of entering input data and requesting output reports containing the process results, and prompts users with questions to help guide them through data inputs to either generate a new network or update an existing network. It offers users a choice of detail in questions: A new user can receive the complete question forms that explain the exact user responses, while an experienced user can receive very brief questions for more rapid progress.

An example of one of the data input (or input windows) screens of the computer installation is shown in Figure 7-5. The basic data required are taken essentially from the planning model.

- *Code:* Each activity has a code at the beginning and end (sometimes noted as *i* and *j; i* is the beginning code, and *j* the ending code). These codes are uniquely numbered so that each activity can be located in the diagram.

- *Description:* A brief phase that describes the work to be done.

- *Duration:* The estimated time for completing each activity or job.

- *WBS:* The work breakdown structure, which is a code used to help locate and group activities of a similar nature, such as same department, task responsibilities of the same person, or elements of the same process. (WBS is one of the most important input features of software, as it will organize the project items in the proper context.)

Optional data that can be entered on the miniscreen include the resources necessary to complete the task and specific *start and finish scheduled* dates that define a job that *cannot start before* a specific date and *cannot finish after* a specific date. As a project control expedient, *actual start* and *actual finish* dates can be entered.

```
                                    Plan:COMINST.3    Proj:COMINST   READY
002 -006
═════════════════════════════════Input Report═══════════════════════════ T ═╗
 ─Name─►      ─────────────Description─────────────► Durtn►  ─W B S ─►  ║
  001 -001    START COMPUTER PROJECT                 Milstn    -        MA
  001 -002    DECIDE ON COMPUTER                        0      A-AA
  001 -006    COMPUTER SPECS, PROCURE COMPUTER       Hammock    -        HB
  002 -003    DETERMINE SITE SPECIFICATIONS             0      A-BA
  002 -006    PROCURE COMPUTER                          61     A-AB
  002 -016    DUMMY                                      0       -
  002 -017    DUMMY                                      0       -
  002 -018    SITE PREPARATION                       Hammock    -        HA
  003 -004    SOLICIT BIDS FOR SITE                     0      A-BB
  004 -005    AWARD SITE CONTRACT                       0      A-BC
  005 -018    PREPARE SITE                              71     A-BD
  006 -006    COMPUTER DELIVERED ON SITE             Milstn    -        MC ▼
  .
╠════════════════════════════════════════════════════════════════════════════
 Calendar►  Org Dur► %Cmplt►
  DAILY       140     57
          Scheduled─►   ─Actual─►
 Start   1/OCT/90      1/OCT/90
 Finsh  12/APR/91       /  /
```

Figure 7-5 Typical input screen.

Other input items to consider: calendars to reflect the proper 5- or 7-day-week schedule, holidays, special days off, and so on; milestones; and hammocks (summarizes paths for management summary reports).

Schedule Reports

Software packages produce any number of reports, and it will be at the discretion of the user to generate those reports that best apply. The nature of the project, as well as project managers and project engineers, will dictate the reports to be used. Of primary concern is that the reports be designed in formats or graphics that are suited for all levels of management to understand and for use in any analysis that project conditions may require. Included among the more useful reports are the following:

- *Milestone report:* lists all milestones (major events) and their calculated (or established) start or finish dates.
- *Work breakdown structure (WBS):* all of the reports can be sorted according to departments, task responsibilities, or any listing of similar nature.
- *Bar chart report:* shows graphically the duration and time frames of the specified work items.
- *Resource availability and utilization reports:* shows in a tabular list or graphically, on a daily, weekly, or monthly basis, each critical resource and its availability and utilization.
- *Hammock (or summary) report:* summarizing a "path" of work tasks that extends from the start of the first task to the finish of the last task.

Milestone Report

The milestone report is an excellent tool for reporting project status in summary form to management and to project personnel. The report displays selected events whose start (or completion) dates are critical to successful completion of the project. The initial milestone report for the Computer Installation Project is show in Figure 7-6.

During the course of the project the milestone report will be updated automatically based on timing changes involving the duration of tasks or activities. Management can note immediately from the updated milestone report if there are major deviations from the original project plan, and if so, should effect any needed corrections or improvements to the plan and schedule to bring it back "on course."

Work Breakdown Structure (WBS)

One of the initial efforts in project planning is to prepare an organizational structure that provides an orderly approach to carrying out the plan. This

```
REPORT TYPE  : MILESTONE SCHEDULE : Normal
TIME NOW DATE:  3/SEP/90    PRODUCED BY: P1211
```

NODE NUMBERS	DESCRIPTION	MILESTONE DATE	FLOAT
001 -001	START COMPUTER PROJECT	3/SEP/90	0
018 -018	COMPLETE SITE PREPARATION	15/APR/91	0
006 -006	COMPUTER DELIVERED ON SITE	15/APR/91	0
014 -014	COMPUTER SYSTEM OPERATIONAL	10/JUN/91	0

- Milestone Report shows start of project on 3 September 1990 and computer systems to become operational on 10 June 1991.

- Site is to be ready by 15 April 1990, when computer installation can start.

- Computer delivery is scheduled for 22 March 1991 (three weeks earlier than completion of site when installation can begin). Authors note: During the planning phase, it is prudent to allow some float time for equipment delivery in the event the suppliers proposed schedule tends to drift. (A "drift" of three weeks can be contained in this project).

Figure 7-6 Milestone report for the Computer Installation Project.

structure can be organized (or sorted) according to department, subcontractor, individuals, or process. In the case of the Computer Installation Project, the schedule adheres to the work breakdown structure (WBS) approach, where the report activities are sorted according to an organization-like structure (see Figure 7-3). This permits the project schedule to be sorted into the major categories. The categories with their respective activities and node number are noted as follows:

COMPUTER SYSTEM

001-002	Decide on computer
002-006	Procure computer
006-011	Install and test computer
011-013	Test computer program
013-014	Operate computer program

SITE PREPARATION

015-003	Determine site specifications
003-004	Solicit bids for site
004-005	Award site contract
005-018	Prepare site

PERSONNEL PROCURE, TRAINING

016-010	Select operating personnel
010-109	Train operating personnel
017-007	Select programming personnel
007-008	Train programming personnel

PROGRAM DEVELOPMENT

008-081	Record layout
081-021	Develop program
009-012	Design forms
012-020	Procure forms

Schedule Reports

The master schedule report (Figure 7-7) provides for a listing of all of the activities (preferably in the work breakdown structure alignment) that will show their starting and ending dates, with any optional dates and associated float times and their durations. This format can be lengthy, depending on the size of the project, but it does allow for a total detailed overview of the project timing.

Used in conjunction with the schedule is a computer-generated bar chart schedule (Figure 7-8) that graphically portrays the related time frames and durations of the project schedule. The bar chart is best understood by dividing the project schedule into its work breakdown structure alignment, which permits better control of the project.

Using the WBS approach, this project is organized into these sections:

1. Computer System (Figure 7-9).
2. Site Preparation (Figure 7-10).
3. Personnel Procure, Training (Figure 7-11).
4. Program Development (Figure 7-12).

It is desirable for a person to be placed in charge of each group that will be responsible for the work effort and the schedule. The assistant project manager uses his or her individual schedule in planning the work. Additional detail developed, which is not shown on the schedule, will need to be developed to effectively control the individual sections.

An example of the additional detail: The project item *Procure Computer* will need to be divided, as the total duration of 25 weeks (125 working days) is too difficult to control as one major activity. This item is to be handled primarily by the supplier, who should provide a detailed schedule of procuring materials, fabricating major sections, assembling computer components, fabricating and

REPORT TYPE : PROJECT SCHEDULE : Normal
TIME NOW DATE: 3/SEP/90   PRODUCED BY: P1211

Name	Description	Early Start	Late Start	Early Finish	Late Finish	Total Float	Org Durtn
001 -002	DECIDE ON COMPUTER	3/SEP/90	3/SEP/90	28/SEP/90	28/SEP/90	0	20
002 -003	DETERMINE SITE SPECIFICATIONS	1/OCT/90	1/OCT/90	5/OCT/90	5/OCT/90	0	5
016 -010	SELECT OPERATING PERSONNEL	1/OCT/90	25/MAR/91	5/OCT/90	29/MAR/91	125	5
017 -007	SELECT PROGRAMMING PERSONNEL	1/OCT/90	19/NOV/90	12/OCT/90	30/NOV/90	35	10
002 -006	PROCURE COMPUTER	1/OCT/90	22/OCT/90	22/MAR/91	12/APR/91	15	125
003 -004	SOLICIT BIDS FOR SITE	8/OCT/90	8/OCT/90	9/NOV/90	9/NOV/90	0	25
010 -019	TRAIN OPERATING PERSONNEL	8/OCT/90	1/APR/91	9/NOV/90	3/MAY/91	125	25
007 -008	TRAIN PROGRAMMING PERSONNEL	15/OCT/90	3/DEC/90	7/DEC/90	25/JAN/91	35	40
004 -005	AWARD SITE CONTRACT	12/NOV/90	12/NOV/90	23/NOV/90	23/NOV/90	0	10
005 -018	PREPARE SITE	26/NOV/90	26/NOV/90	12/APR/91	12/APR/91	0	100
008 -081	RECORD LAYOUT	10/DEC/90	28/JAN/91	21/DEC/90	8/FEB/91	35	10
009 -012	DESIGN FORMS	24/DEC/90	29/APR/91	4/JAN/91	10/MAY/91	90	10
081 -021	DEVELOP PROGRAM	24/DEC/90	11/FEB/91	15/MAR/91	3/MAY/91	35	60
012 -020	FORMS PROCUREMENT	7/JAN/91	13/MAY/91	18/JAN/91	24/MAY/91	90	10
006 -011	INSTALL AND TEST COMPUTER SYSTEM	15/APR/91	15/APR/91	3/MAY/91	3/MAY/91	0	15
011 -013	TEST COMPUTER PROGRAM	6/MAY/91	6/MAY/91	24/MAY/91	24/MAY/91	0	15
013 -014	OPERATE COMPUTER PROGRAM	27/MAY/91	27/MAY/91	7/JUN/91	7/JUN/91	0	10

• COMPUTER (Comp): From a timing standpoint, this category is the most critical. Four of the five project activities are on the critical path, and the fifth, Procure computer (with just three weeks of float time) can "drift" by three weeks without causing a potential delay in the program. Careful surveillance of the computer supplier should be an essential part of the planning process.

• SITE (Site): This category is essentially as critical as the COMPUTER category, insofar as all of the project items are on the critical path. Experienced and qualified personnel need to assume responsibility to ensure that the site will be ready in time to receive the computer equipment and hardware. Continued surveillance of the suppliers progress is most essential.

• PERSONNEL (Pers) and PROGRAM DEVELOPMENT (Prog Dev): As these categories have adequate float time, the persons responsible can prepare a more definitive schedule for implementation.

Figure 7-7 Project schedule for the Computer Installation Project.

assembling controls, and shipping. An owner representative will visit suppliers periodically to monitor the status of this item.

Analysis of Results

At this point there is a review of the project plan and schedule, observing the project milestones and project durations to determine whether all of the objectives can be met. If the project duration is not satisfactory, one of the first

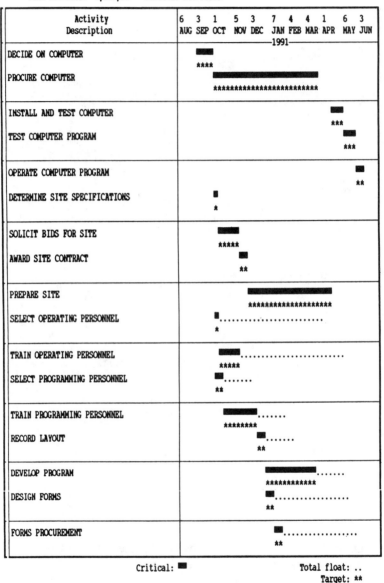

Figure 7-8 Bar chart schedule.

REPORT TYPE : COMPUTER SYSTEM : Normal
TIME NOW DATE: 3/SEP/90   PRODUCED BY: P1211

COMPUTER INSTALLATION PROJECT

Node Numbers	Description	Early Start	Late Start	Early Finish	Late Finish	Duration	Float
001 -002	DECIDE ON COMPUTER	3/SEP/90	3/SEP/90	28/SEP/90	28/SEP/90	20	0
002 -006	PROCURE COMPUTER	1/OCT/90	22/OCT/90	22/MAR/91	12/APR/91	125	15
006 -011	INSTALL AND TEST COMPUTER	15/APR/91	15/APR/91	3/MAY/91	3/MAY/91	15	0
011 -013	TEST COMPUTER PROGRAM	6/MAY/91	6/MAY/91	24/MAY/91	24/MAY/91	15	0
013 -014	OPERATE COMPUTER PROGRAM	27/MAY/91	27/MAY/91	7/JUN/91	7/JUN/91	10	0

> - The Computer System section will undoubtedly require the most direction, as all but one of the activities included within this section are critical to the project -- any scheduled "drift" in any of the individual activities is a potential delay in the completion of the total project.
>
> - The computer supplier has specified a six month delivery (125 days duration based on a five day week schedule). For a computer of this type the three weeks float time can be treated as if this item is also on the "critical path." As there are strong possibilities that equipment supplier schedules "drift", the Owner's representative needs to pay special attention to the suppliers schedule, making certain that the supplier provides a reliable schedule at the time that the contract is signed.
>
> - There were some "after thoughts" that the activity, Install and Test Computer has adequate time for its duration (15 days). To protect this schedule from the possibility of some delay, there should be assurance that the computer will arrive on the site as originally planned.

Figure 7-9 Computer system schedule.

steps is to reevaluate the time requirements of jobs on the critical path to determine which of them can be reduced most readily. Another approach is to replan the way in which the work will be done in order to shorten the critical path and meet deadline requirements.

Having replanned the diagram and reevaluated the durations, a second computer run is made before initiating the project. It is likely that replanning and retiming of only a small part of the network will be needed, comprising for the most part activities on the critical path that will have to be changed. Most of the original inputs will still be valid, and making input corrections is not nearly as difficult using the new software packages as it was preparing punchcards before the era of personal computers and the interactive input approach. At the present time, almost all project management software packages use the interactive method of processing project input data.

REPORT TYPE : SITE PREPARATION : Normal
TIME NOW DATE: 3/SEP/90   PRODUCED BY: P1211

Node Numbers	Description	Early Start	Late Start	Early Finish	Late Finish	Duration	Float
002 -003	DETERMINE SITE SPECIFICATIONS	1/OCT/90	1/OCT/90	5/OCT/90	5/OCT/90	5	0
003 -004	SOLICIT BIDS FOR SITE	8/OCT/90	8/OCT/90	9/NOV/90	9/NOV/90	25	0
004 -005	AWARD SITE CONTRACT	12/NOV/90	12/NOV/90	23/NOV/90	23/NOV/90	10	0
005 -018	PREPARE SITE	26/NOV/90	26/NOV/90	12/APR/91	12/APR/91	100	0

> • Activities associated with Site Preparation are
> critical -- all of their schedules must be met
> to avoid delays in the start-up of the computer
> system.
>
> • Determine site specifications and Solicit bids
> for site are the most important to maintaining
> a reliable schedule. They establish the
> momentum for a timely completion.
>
> • As this plan and schedule was developed prior
> to the start of the project, consideration
> should be given during the preparation of the
> specifications on reducing the Prepare Site
> duration time from 20 weeks (based upon a 5
> day week schedule) to about 16 weeks. An
> informal review with the contractors during
> the solicit bid period may determine the
> validity of such a decision.

Figure 7-10 Site preparation schedule.

Project Control

A satisfactory plan and schedule should be completed about the time the project is to be implemented. The early start schedule is the one usually used to get the project rolling. It will generally be found that soon after work begins on a project, one or more of the assumptions underlying the original plan must be changed. It is possible that jobs in the project may require more or less time than was anticipated, or a supplier may not keep their initial delivery promise. Also, the assumption that one activity may follow another could prove to be invalid. Thus as the project progresses, it is important to update both the time estimates and the network logic to indicate any changes in the plan.

The total project control process requires systematic appraisal on a periodic basis. A monthly report is prepared for normal projects, and projects in "trouble" may require more frequent formal appraisal. Projects known to be "safe" may only be reviewed every 2 months or so.

The main purpose of project control is to keep the overall project on time and within budget. Use of the early-warning system technique, which allows intervention at an early date to begin correction procedures, is the standard operating procedure. These are four steps in this technique:

COMPUTER INSTALLATION PROJECT

REPORT TYPE : PERSONNEL PROCURE, TRAINING : Normal
TIME NOW DATE: 3/SEP/90   PRODUCED BY: P1211

Node Numbers	Description	Early Start	Late Start	Early Finish	Late Finish	Duration	Float
016 -010	SELECT OPERATING PERSONNEL	1/OCT/90	25/MAR/91	5/OCT/90	29/MAR/91	5	125
010 -019	TRAIN OPERATING PERSONNEL	8/OCT/90	1/APR/91	9/NOV/90	3/MAY/91	25	125
017 -007	SELECT PROGRAMMING PERSONNEL	1/OCT/90	19/NOV/90	12/OCT/90	30/NOV/90	10	35
007 -008	TRAIN PROGRAMMING PERSONNEL	15/OCT/90	3/DEC/90	7/DEC/90	25/JAN/91	40	35

- The early start and early finish schedule for the Personnel procure and training has a wide range of timing flexibility. (For example, the start time for the activity, Select operating personnel, has a start time range from 1/Oct/90 to 25/Mar/91 -- no delay in program will be experienced if this activity is started within this period.)

- Prudent project management practice suggest that activities should start as early as possible to provide "cushions" for any potential delays. (It is conceivable that the search for qualified personnel through the newspaper media in various cities may take longer than expected). Contingencies of various possibilities support the start of activities as early as possible.

- However, bringing personnel "on board" at any early date may incur added costs before they are actually needed. Therefore, planning the float times, especially the operating personnel is important at the early stages of the project.

Figure 7-11 Personnel procure, training schedule.

1. Monitor the status of each activity.
2. Assess the status.
3. Resolve concerns.
4. Communicate.

Using a computer makes it possible to keep this technique effective. It needs to be emphasized that the technique does not replace human thought and analysis, but does provide support.

Monitoring consists of reviewing start and finish dates and the duration of each current project activity. One technique employed is to mark up the most recent computer run and the new timing data, which then become the source document to input the computer program. Figure 7-13 is a typical worksheet showing the results of meetings with the project principals.

Assessing the project status begins with a new computer run that shows the actual start and finish dates and the revised durations that were entered.

COMPUTER INSTALLATION PROJECT

REPORT TYPE   : PROGRAM DEVELOPMENT : Normal
TIME NOW DATE: 3/SEP/90   PRODUCED BY: P1211

Node Numbers	Description	Early Start	Late Start	Early Finish	Late Finish	Duration	Float
008 -081	RECORD LAYOUT	10/DEC/90	28/JAN/91	21/DEC/90	8/FEB/91	10	35
081 -021	DEVELOP PROGRAM	24/DEC/90	11/FEB/91	15/MAR/91	3/MAY/91	60	35
009 -012	DESIGN FORMS	24/DEC/90	29/APR/91	4/JAN/91	10/MAY/91	10	90
012 -020	FORMS PROCUREMENT	7/JAN/91	13/MAY/91	18/JAN/91	24/MAY/91	10	90

- Activities associated with Program Development, whose schedules are dependent on the outcome of the programming and operating personnel schedules, also have relatively large float times. While monitoring these activities will not require the degree of attention of the more critical activities, there is opportunity for these activities to be managed effectively to the benefit of the total project.

- For example, additional time can be allotted to designing and procuring forms to acquire better pricing proposals.

- Schedules for the programming personnel and program development sections should be monitored by the same person for the most effective management. The large float times associated with the activities in these sections still require some degree of coordination with the little or no float activities where they are dependent on their start and/or finish dates.

Figure 7-12   Program development schedule.

One appropriate computer run (shown in Figure 7-14) prioritizes the critical items on which further action needs to be taken to return the project to its original schedule.

Another computer printout used to interpret and assess status is the milestone report. The updated milestone report (Figure 7-15) is compared with the last milestone report (Figure 7-6) to note trends and is valuable in resolving potential concerns.

Resolving concerns uses the project management software to produce simulations. One of the expedients that project personnel can apply are "what-if" exercises. These types of exercises are made possible with the software that is now available (request software vendors to demonstrate). Exercises of this nature assist in an analysis that may make it possible to restore the original project objectives.

Communicating the status to management and to personnel associated with a project can become effective through formal reporting. The project status report (explained in more detail in Chapter 4) informs management, who, by reviewing the report, can make an appraisal of the project. The report highlights critical areas to management and other project personnel, so that

COMPUTER INSTALLATION PROJECT

REPORT TYPE : PROJECT WORKSHEET : Normal
TIME NOW DATE: 20/JAN/91   PRODUCED BY: P1211

Node Numbers	Description	Early Start	Late Start	Actual Start	Early Finish	Late Finish	Actual Finish	Duration	Float
001 -002	DECIDE ON COMPUTER	3/SEP/90	3/SEP/90	3/SEP/90	28/SEP/90	28/SEP/90	28/SEP/90	20	0
002 -006	PROCURE COMPUTER	1/OCT/90	1/OCT/90	1/OCT/90	15/APR/91	12/APR/91	/ /	140	-1
006 -011	INSTALL AND TEST COMPUTER SYSTEM	30/APR/91	15/APR/91	/ /	20/MAY/91	3/MAY/91	/ /	15	-11
011 -013	TEST COMPUTER PROGRAM	21/MAY/91	6/MAY/91	/ /	10/JUN/91	24/MAY/91	/ /	15	-11
013 -014	OPERATE COMPUTER PROGRAM	11/JUN/91	27/MAY/91	/ /	24/JUN/91	7/JUN/91	/ /	10	-11
002 -003	DETERMINE SITE SPECIFICATIONS	1/OCT/90	1/OCT/90	1/OCT/90	5/OCT/90	5/OCT/90	5/OCT/90	5	0
003 -004	SOLICIT BIDS FOR SITE	8/OCT/90	8/OCT/90	8/OCT/90	9/NOV/90	9/NOV/90	9/NOV/90	25	0
004 -005	AWARD SITE CONTRACT	12/NOV/90	12/NOV/90	12/NOV/90	23/NOV/90	23/NOV/90	23/NOV/90	10	0
005 -018	PREPARE SITE	26/NOV/90	26/NOV/90	26/NOV/90	29/APR/91	12/APR/91	/ /	110	-11
016 -010	SELECT OPERATING PERSONNEL	1/OCT/90	1/OCT/90	1/OCT/90	5/OCT/90	5/OCT/90	5/OCT/90	5	0
010 -019	TRAIN OPERATING PERSONNEL	8/OCT/90	8/OCT/90	8/OCT/90	9/NOV/90	3/MAY/91	9/NOV/90	25	125
017 -007	SELECT PROGRAMMING PERSONNEL	1/OCT/90	1/OCT/90	1/OCT/90	12/OCT/90	12/OCT/90	12/OCT/90	10	0
007 -008	TRAIN PROGRAMMING PERSONNEL	15/OCT/90	15/OCT/90	15/OCT/90	7/DEC/90	7/DEC/90	7/DEC/90	40	0
008 -081	RECORD LAYOUT	10/DEC/90	10/DEC/90	10/DEC/90	21/DEC/90	21/DEC/90	21/DEC/90	10	0
081 -021	DEVELOP PROGRAM	24/DEC/90	24/DEC/90	24/DEC/90	26/APR/91	3/MAY/91	/ /	70	5
009 -012	DESIGN FORMS	24/DEC/90	24/DEC/90	24/DEC/90	4/JAN/91	4/JAN/91	4/JAN/91	10	0
012 -020	FORMS PROCUREMENT	7/JAN/91	7/JAN/91	7/JAN/91	18/JAN/91	24/MAY/91	18/JAN/91	10	90

Figure 7-13 Project worksheet (January 20, 1991). (General note: Where negative figures are shown under the *Float* column, depicting a behind schedule situation, behind schedule dates will be shown in either (or both) the *Early Start* and *Early Finish* columns.)

137

COMPUTER INSTALLATION PROJECT

REPORT TYPE : CRITICAL ITEMS : Normal
TIME NOW DATE: 20/JAN/91   PRODUCED BY: P1211

Name	Description	Early Start	Late Start	Early Finish	Late Finish	Duration	Float
005 -018	PREPARE SITE	26/NOV/90	26/NOV/90	29/APR/91	12/APR/91	110	-11
006 -011	INSTALL AND TEST COMPUTER SYSTEM	30/APR/91	15/APR/91	20/MAY/91	3/MAY/91	15	-11
011 -013	TEST COMPUTER PROGRAM	21/MAY/91	6/MAY/91	10/JUN/91	24/MAY/91	15	-11
013 -014	OPERATE COMPUTER PROGRAM	11/JUN/91	27/MAY/91	24/JUN/91	7/JUN/91	10	-11
009 -012	DESIGN FORMS	24/DEC/90	24/DEC/90	18/JAN/91	4/JAN/91	10	-10
002 -006	PROCURE COMPUTER	1/OCT/90	1/OCT/90	15/APR/91	12/APR/91	140	-1
001 -002	DECIDE ON COMPUTER	3/SEP/90	3/SEP/90	28/SEP/90	28/SEP/90	20	0
002 -003	DETERMINE SITE SPECIFICATIONS	1/OCT/90	1/OCT/90	5/OCT/90	5/OCT/90	5	0
003 -004	SOLICIT BIDS FOR SITE	8/OCT/90	8/OCT/90	9/NOV/90	9/NOV/90	25	0
004 -005	AWARD SITE CONTRACT	12/NOV/90	12/NOV/90	23/NOV/90	23/NOV/90	10	0
007 -008	TRAIN PROGRAMMING PERSONNEL	15/OCT/90	15/OCT/90	7/DEC/90	7/DEC/90	40	0
008 -081	RECORD LAYOUT	10/DEC/90	10/DEC/90	21/DEC/90	21/DEC/90	10	0
016 -010	SELECT OPERATING PERSONNEL	1/OCT/90	1/OCT/90	5/OCT/90	5/OCT/90	5	0
017 -007	SELECT PROGRAMMING PERSONNEL	1/OCT/90	1/OCT/90	12/OCT/90	12/OCT/90	10	0
081 -021	DEVELOP PROGRAM	24/DEC/90	24/DEC/90	29/MAR/91	3/MAY/91	70	25
012 -020	FORMS PROCUREMENT	21/JAN/91	7/JAN/91	18/JAN/91	24/MAY/91	10	90
010 -019	TRAIN OPERATING PERSONNEL	8/OCT/90	8/OCT/90	9/NOV/90	3/MAY/91	25	125

Figure 7-14 Critical items. (General note: Where negative figures are shown under the *Float* column, depicting a behind schedule situation, behind schedule dates will be shown in either (or both) the *Early Start* and *Early Finish* columns.)

REPORT TYPE : (UPDATED) MILESTONE SCHEDULE : Normal
TIME NOW DATE: 20/JAN/91   PRODUCED BY: P1211

Description	Milestone Dates	Float
START COMPUTER PROJECT	3/SEP/90	0
COMPLETE SITE PREPARATION	30/APR/91	-11
COMPUTER DELIVERED ON SITE	16/APR/91	-1
COMPUTER SYSTEM OPERATIONAL	25/JUN/91	-10

Figure 7-15   Updated milestone schedule.

they can objectively evaluate the problem areas. Figure 7-16 is a sample of a management summary report. Sometimes called a "hammock report," the summary can also be reported in tabular form. It can show additional information for each "hammock," including optional schedule dates and float times. Figure 7-17 shows a summary report that allows for an effective overview of the Computer Installation Project, avoiding the need to review every activity of the project.

Resource Planning

Another area in which computer software has an advantage over the manual approach is in developing personnel work loads for a project. Figures

REPORT TYPE   : MANAGEMENT BAR CHART SUMMARY : Target
TIME NOW DATE: 20/JAN/91   PRODUCED BY: P1211

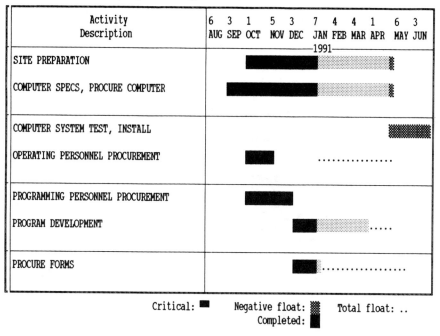

Figure **7-16**   Management summary report.

REPORT TYPE   : MANAGEMENT SUMMARY (HAMMOCK) : Normal
TIME NOW DATE: 20/JAN/91   PRODUCED BY: P1211

Description	Early Start	Late Start	Early Finish	Late Finish	Float
SITE PREPARATION	1/OCT/90	4/JAN/91	29/APR/91	12/APR/91	-11
COMPUTER SPECS, PROCURE COMPUTER	3/SEP/90	4/JAN/91	29/APR/91	12/APR/91	-11
COMPUTER SYSTEM TEST, INSTALL	30/APR/91	16/APR/91	24/JUN/91	10/JUN/91	-10
OPERATING PERSONNEL PROCUREMENT	1/OCT/90	6/MAY/91	9/NOV/90	3/MAY/91	125
PROGRAMMING PERSONNEL PROCUREMENT	1/OCT/90	10/DEC/90	7/DEC/90	7/DEC/90	0
PROGRAM DEVELOPMENT	10/DEC/90	25/FEB/91	29/MAR/91	3/MAY/91	25
PROCURE FORMS	10/DEC/90	27/MAY/91	18/JAN/91	24/MAY/91	90

Figure **7-17**   Tabulated management summary ("hammock") report.

7-18 and 7-19 show the initial planning effort for allocating analysts and programming resources prior to leveling. Leveling in connection with schedule changes is discussed in Chapter 6. Leveling exercises for this project are included in the Appendix.

Implementation of the Project

It would be presumptuous to think that all there is to running a project is following the progress of the project and advising on critical situations. The

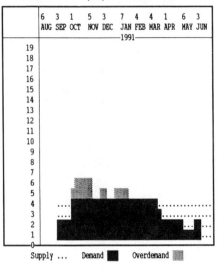

°This graphic report is a weekly summary of the analysts supply
and demand starting on 1 September 1990 and completing on 10
June 1991. The reporting period of 40 weeks is noted
graphically by each vertical line on the histogram representing
one week.

°There is a constant supply of four analysts over the total
project duration of 40 weeks.

°On seven of the 40 weeks the requirements exceed the supply
by one; and for another five weeks there is an "overdemand"
of two additional analysts that will be required.

°If overtime is permitted by the organization, and there is
a normal 40 hour work week, then each of the four analysts
assigned to this project could work an additional 10 hours
per week for the seven week period, and an additional 20 hours
per week for the five week period.

°When overtime may not be feasible, and no more analysts can
be assigned to this project, there are two alternatives:

    -Utililizing the available float times, "fit" the four
    analysts within the project timing.

    -If there are still more analysts required than available,
    then the project timing will need to be extended,

Figure 7-18   Analyst demand and supply.

REPORT TYPE   : PROGRAMMER SUPPLY AND DEMAND : Histogram-PROGRAMMER
TIME NOW DATE: 3/SEP/90   PRODUCED BY: P1211

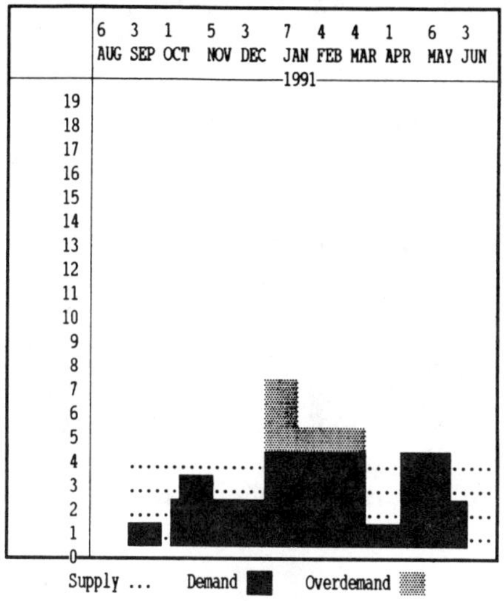

```
        6  3  1   5  3   7  4  4  1   6  3
        AUG SEP OCT  NOV DEC  JAN FEB MAR APR  MAY JUN
                            ─────1991─────
   19
   18
   17
   16
   15
   14
   13
   12
   11
   10
    9
    8
    7
    6
    5
    4
    3
    2
    1
    0
    Supply ...     Demand ■■■     Overdemand ▒▒▒
```

-This graphic report, showing a weekly summary of the programmers
 supply and demand, with a 1 September 1990 start date and a
 10 June 1991 project completion date. The summary is based
 upon the early start schedule.

-Two programmers were made available at the start of the project
 for this project. However, the programmer demand reaches
 proportions beyond use of overtime to overcome so that either
 additional programmers are required or the duration of the
 project needs to be extended.

-On four of the forty weeks duration of the project five
 additional programmers are needed daily; for eight weeks, three
 additional programmers daily; and for four weeks, one additional
 programmer daily.

-Programmer requirements are more critical than the analyst
 requirements, therefore "fit" or leveling exercise should
 be conducted to determine the length of the project based upon
 the allotted supply of the programmers to the project.

Figure 7-19   Programmer demand and supply.

day-by-day operation concerns itself with managing all elements associated
with project management. These elements are identified as follows:

1. *Managing human resources* concentrates on the people in an organization.

2. *Managing costs* relates to accumulating, organizing, and analyzing data
   for making cost-related decisions.

3. *Managing time* consists of planning, scheduling, and controlling (including monitoring) a project to achieve the time objectives of the project.

4. *Managing communications* concerns itself with information flow among members of the project team.

5. *Managing the scope* of a project involves controlling the project based on an understanding of the aims, goals, and objectives of the sponsor of the project.

6. *Managing the quality* of a project assures the production of a quality product or result upon completion of the project.

(Managing procurement and managing risk are two additional elements that are also to be considered in the daily operations. Discussion of these two elements and managing quality have not been included in this book, as they tend to be more advanced subjects.)

Time management and cost management may be the very essence of project management, with cost management considered to be the primary function.

SUMMARY

There are many benefits to be derived from using a computer for network planning. With a computer it is possible to:

1. Apply the network planning method very rapidly to large projects that would require many hours of manual calculation.

2. Handle accurately a project in which manual calculations would be subject to errors.

3. Make updating runs as often as desired without an excessive expenditure of time.

4. Print in useful and readable form results of the computer analysis.

5. Undertake time/cost trade-off and resource-leveling analysis that would be difficult and extremely time consuming to do manually.

8
PERT/Time

PERT (Project Evaluation Review Technique) is a technique that can be used to plan, schedule, and control activities that must be completed to finish a project. Developed in the late 1950s for the Special Projects Office of the Navy Bureau of Ordnance, PERT was used initially to plan and coordinate the work of some 3,000 contractors and agencies for the Polaris missile program. The use of PERT is credited with advancing successful completion of the Polaris program more than 2 years.

PERT and CPM are similar in concept: both use arrow diagrams, calculate critical paths and floats, and use computers to carry out detailed calculations. The two techniques differ in several significant details:

1. PERT is "event-oriented."
2. PERT uses three time estimates for each activity.
3. PERT calculates the probability of meeting a scheduled date.

Definition of Terms

Project: a network of activities and events having well-defined starting and ending points.

The material in Chapter 8 is taken from *Philco Manual TM-19 PERT Systems,* April 1962, and *TM-30 PERT III Systems.*

Event: the starting and/or ending point of an activity. Events designate either specific accomplishments or points at which programs start.

Activity: a time-consuming element of a project which is defined by a predecessor and successor event. An activity cannot start until its predecessor event is completed.

Slack float: the time interval until the completion date of the end event.

ESTABLISHING THE NETWORK

After setting objectives, the first step in planning a PERT system is to determine the activities that must be accomplished. When qualified individuals have determined what activities are required, an arrow diagram is developed that accurately depicts the interrelations—which activity must come first, second, and so on, and which activities can be performed concurrently with others.

Each activity in the project is represented by an arrow and label on the diagram. It should be noted, however, that, in contrast with CPM networks, in which activities are labeled, in the PERT network the nodes or events are labeled. For example, Event 1 in a network might have the title "Contract Awarded"; Event 2, "Start Computer Design"; and so on.

OBTAINING TIME ESTIMATES

Time estimates are obtained from a person who is very familiar with the project. The three time estimates—optimistic, most likely, and pessimistic—for each activity are used to offset the bias that is usually present in one time estimate. The range of the time estimates also gives some indication of the scheduling risk involved; for example, a wide spread between the estimates shows considerable uncertainty about the time actually required to accomplish an activity.

Generally, PERT programs require that time estimates be converted into calendar dates, because military users have found this method convenient for their purposes. The computer program usually translates calendar days into weeks and fractions of weeks.

Calendar scheduled dates, or contractual dates, may also be recorded in the PERT diagram. These scheduled dates tend to complicate the job of obtaining realistic time estimates, in that persons tend to give estimates that fit into the time available. This effect may be minimized by securing time estimates for various activities independently, by aggregating estimates for small sections of the network separately, and by releasing scheduled dates only after the detailed time estimates have been obtained. The dates derived by looking at the work content can then be compared to the schedule to see whether discrepancies exist.

Most Likely Time (or Normal Time)

The first time estimate requested is the most likely time. It is the time that would be most frequently required if the activity were repeated many times under similar conditions. It is also the job time estimate that would be used in the critical path method. In a frequency distribution, the most likely time would be the *mode* of the distribution. On the histogram (Figure 8-1), the time period 5 has the greatest frequency (6).

Optimistic Time

The shortest possible time required for completing an activity is the optimistic time. Here it is assumed that everything goes as planned: deliveries of material occur on schedule, machines operate without major breakdowns, personnel perform work within work standards, and the like. The time period 3 on the frequency distribution (Figure 8-1) would represent the optimistic time.

Pessimistic Time

The maximum possible time required to complete an activity is termed the pessimistic time. This is the time required for doing something if everything goes wrong simultaneously—in short, the worst possible situation. On the histogram (Figure 8-1) the pessimistic time is 9.

SAMPLE PERT NETWORK

The three time estimates obtained for each activity in the project can be recorded on the PERT network to aid in the planning stage. For illustration purposes, a sample PERT network is shown in Figure 8-2 with the time estimates indicated. There are six events (figures in circles) in the network: 1, 2, 3, 4, 5, and 6. The events are connected by arrows representing eight activities or jobs to which the sets of three time estimates have been recorded:

Activity	Time Estimates (weeks)
1,2	1–3–8
2,3	3–6–9
2,4	1–2–3
etc.	

This sample network with six events and eight activities will be used to explain the principles involved in making PERT timing calculations. For a network with, say, more than 100 activities, a computer is used to perform the timing calculations.

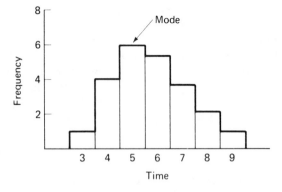

Figure 8-1   Frequency distribution chart.

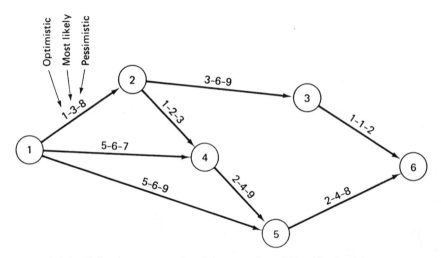

<u>Labels.</u> Following are examples of the type of word identifications that
are listed on a PERT diagram inside each node and on each activity arrow:

Node	Designation	Activity	Designation
1	Start design	1, 2	Electrical design
2	Finish electrical design	2, 3	Order components
3	Component delivery	3, 6	Test components
Etc.		Etc.	

Figure 8-2   PERT network time estimates.

CALCULATING THE EXPECTED (MEAN) TIME AND VARIANCE

With the three time estimates, the expected (mean) time (t_e) and variance (σ_{te})^2 of an activity performance time can be derived. Statisticians have provided formulas for calculating t_e and (σ_{te})^2 for an activity.

Distribution of Time Estimates

The basic factors used in the formulas—the three time estimates—can be translated into a distribution curve, which may resemble one of the curves shown in Figure 8-3.

The location of the time estimates on the time scale determines the type of curve that might represent the distribution—the spread of a (optimistic time) and b (pessimistic time) and the location of m (most likely time) within the spread. To illustrate, the histogram in Figure 8-4(a) (with a distribution curve added) depicts the frequency of occurrences for different time intervals at which an activity might be completed if it were performed many times.

In Figure 8-4(b) the three time estimates represented by the histogram are located on the distribution curve. It is assumed that the curve has only one peak—the most likely time for completion. The m on this peak represents the completion date that has the greatest probability of occurring, while the low points, a and b, indicate dates that have a small chance of being realized.

Formula for Expected (Mean) Time

The three time estimates for each activity are used in calculating a single weighted average or mean called the *expected time* (t_e) of the activity. The

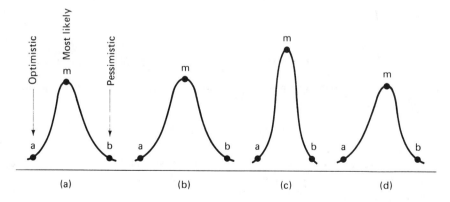

Figure 8-3 Sample distribution curves.

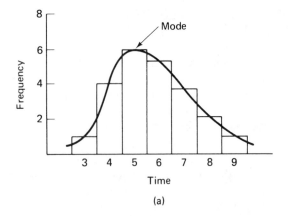

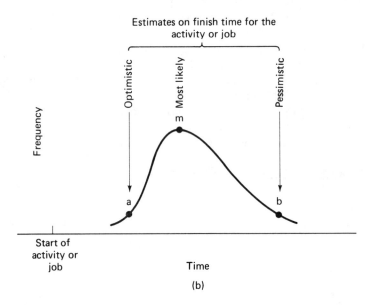

Figure 8-4 (a) Histogram; (b) Distribution curve of estimates on finish time for the activity.

expected time is derived from the optimistic, most likely, and pessimistic times
by this formula:

$$t_e = \frac{a + 4m + b}{6}$$

The symbols a, m, and b represent the three time estimates: a for the optimistic;
b, the most likely (or normal); and c, the pessimistic. The result of the formula
is a weighted average with two-thirds of the weight given to the most likely
(normal) time; one-sixth, to the optimistic; and one-sixth, to the pessimistic.

Example: The expected times for the activities on the sample PERT network are derived from the specified sets of time estimates (Figure 8-2) by applying the formula for t_e.

| | Time Estimates | | | | | Expected |
Activity	Optimistic (a)	Most likely (m)	Pessimistic (b)	Calculation	=	Time, t_e (weeks)
1,2	1	3	8	$\dfrac{1 + 4(3) + 8}{6}$	=	3.5
1,5	5	6	9	$\dfrac{5 + 4(6) + 9}{6}$	=	6.3
2,3	3	6	9	$\dfrac{3 + 4(6) + 9}{6}$	=	6.0
etc.						

All of the expected times are recorded on the network shown in Figure 8-5. Figure 8-6 indicates the location of t_e, the expected time, on the four distribution curves shown previously in Figure 8-3.

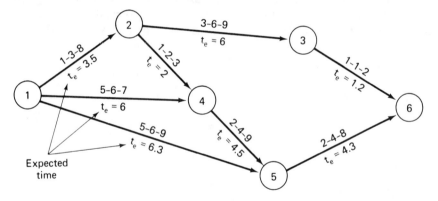

Figure 8-5 Expected time for work items in network.

Formula for Variance

In the PERT system, *variance* is a term that describes the uncertainty associated with how much time will be required to accomplish an activity or the entire project. If the variance is large (which indicates that the optimistic and pessimistic estimates are far apart), there is great uncertainty as to when the activity will be completed. On the other hand, a small variance indicates very little uncertainty. By inspection one can guess that the expected time for Figure 8-6(a) will have a larger variance than the expected time for 8-6(c). In statistical terms, the variance is equal to the standard deviation squared:

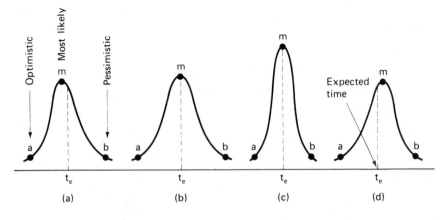

Figure 8-6 Examples of variance of expected times.

$$\text{Variance} = (\sigma_{t_e})^2$$

The *standard deviation* (σ) is a measure of the spread of a distribution. It is the root mean square of the deviations of the various items from their average. For the purposes here, the standard deviation can be approximated as being equal to one-sixth of the range. To illustrate the calculation of the standard deviation and the variance, consider an activity with time estimates of 3-5-10 (Figure 8-7).

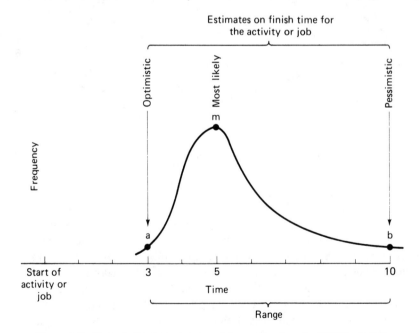

Figure 8-7 Normal distribution curve of time estimates of a PERT activity.

	Formula	Calculation
The range is	$b - a$	$10 - 3 = 7$
The standard deviation (σ) is approximately	$\dfrac{b - a}{6}$	$\dfrac{10 - 3}{6} = \dfrac{7}{6} = 1.17$
The variance $(\sigma_{t_e})^2$ is	$\left(\dfrac{b - a}{6}\right)^2$	$(1.17)^2 = 1.37$

The computation of variance for the activities in the sample PERT network is shown in the following:

Activity	Time — Optimistic	Time — Pessimistic	Calculation	Variance $(\sigma_{t_e})^2$
1,2	1	8	$\left(\dfrac{8 - 1}{6}\right)^2 = \left(\dfrac{7}{6}\right)^2 = (1.17)^2$ =	1.37
1,4	5	7	$\left(\dfrac{7 - 5}{6}\right)^2 = \left(\dfrac{2}{6}\right)^2 = \dfrac{1}{9}$ =	0.11
1,5	5	9	$\left(\dfrac{9 - 5}{6}\right)^2 = \left(\dfrac{4}{6}\right)^2 = \dfrac{4}{9}$ =	0.44

The sample PERT network, Figure 8-8, shows the variances for all eight activities in the project.

CALCULATING EARLIEST EXPECTED TIME, LATEST ALLOWABLE TIME, AND SLACK

With the PERT system, the interest is primarily in events—the starts and finishes of activities. An activity represents work, whereas an event signifies a point in time.

The earliest expected and latest allowable times for an event, and slack (the difference between the two), are a part of the basic data required in scheduling and controlling a PERT project.

Earliest Expected Time

For each event in the network, an *earliest expected time of completion* (T_E) is computed. The T_E for an event is the total obtained by adding the t_e's

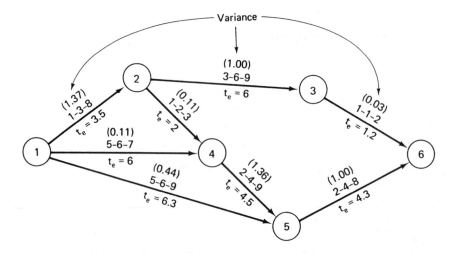

Figure 8-8 Time variances for PERT activities.

(expected times) for the activities on the longest path (in terms of time) that leads to the event. The computation of earliest expected times begins with the first event and continues with the second, and so on, to the end of the project.

The Formula: The first event is established as time zero. For all other events, the computation for determining T_E consists of the following:

$$T_E \text{ for Event } X =$$

$$\left(\begin{array}{c} T_E \text{ for event immediately} \\ \text{preceeding Event } X \end{array}\right) - \left(\begin{array}{c} t_e \text{ for activity immediately} \\ \text{preceding Event } X \end{array}\right)$$

If more than one path leads into Event X, perform the computations for all such paths and use the *largest* total as the T_E for Event X.

Example: The PERT network for the sample project (Figure 8-9) is used to illustrate the calculation of T_E.

		Earliest Expected Time, T_E
Event 1	The earliest expected time for the first event in the project is established as Time 0.	0
Event 2	The longest path—and the only one—leading to Event 2 consists of Activity 1,2. This activity has an expected time of 3.5 weeks.	3.5
	$\dfrac{T_E \text{ for Event 1}}{0} + \dfrac{t_e \text{ for Activity 1,2}}{3.5} = 3.5$	

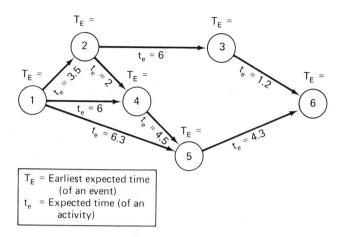

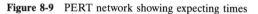

Figure 8-9 PERT network showing expecting times

	Earliest Expected Time, T_E
Event 3 The only path leading to Event 3 is via Activity 2,3. To the T_E for Event 2, add the t_e for Activity 2,3: $$3.5 + 6 = 9.5$$ (see Figure 8-10).	9.5
Event 4 Two paths lead to Event 4—(a) via Activity 2,3; and (b) via Activity 1,4: (a) $\dfrac{T_E \text{ for Event 2}}{3.5} + \dfrac{t_e \text{ for Activity 2,4}}{2} = 5.5$ (b) $\dfrac{T_E \text{ for Event 1}}{0} + \dfrac{t_e \text{ for Activity 1,4}}{6} = 6$ Use the *largest* time factor (see Figure 8-11).	6
Event 5 Two paths lead to Event 5—(a) via Activity 4,5; and (b) via Activity 1,5: (a) $\dfrac{T_E \text{ for Event 4}}{6} + \dfrac{t_e \text{ for Activity 4,5}}{4.5} = 10.5$ (b) $\dfrac{T_E \text{ for Event 1}}{0} + \dfrac{t_e \text{ for Activity 1,5}}{6.3} = 6.3$ Use the *largest* time factor.	10.5
Event 6 Two paths lead to Event 6—(a) via Activity 3,6; and (b) via Activity 5,6: (a) $\dfrac{T_E \text{ for Event 3}}{9.5} + \dfrac{t_e \text{ for Activity 3,6}}{1.2} = 10.7$ (b) $\dfrac{T_E \text{ for Event 5}}{10.5} + \dfrac{t_e \text{ for Activity 5,6}}{4.3} = 14.8$ Use the *largest* time factor.	14.8

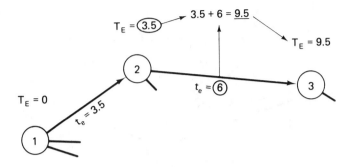

Figure 8-10 T_E for Event 3.

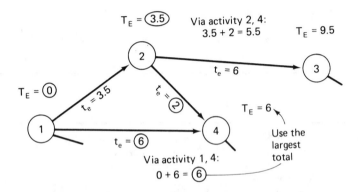

Figure 8-11 Earliest expected time (T_E) for Event 4.

All of the earliest expected times for the events in the sample project are shown on the network in Figure 8-12.

Latest Allowable Time

The *latest allowable time* for an event is both:

• The latest time for completion of the activities that immediately precede the event.

• The latest allowable starting time for the most critical activity that immediately succeeds the event.

The calculation begins with the last event in the project and continues back in reverse sequence through the network to the first event.

The Formula: The T_L for the last event is established either (1) as the time already determined as T_E for that event (the duration of the project if the original time estimates are used in accomplishing the project), or (2) as some

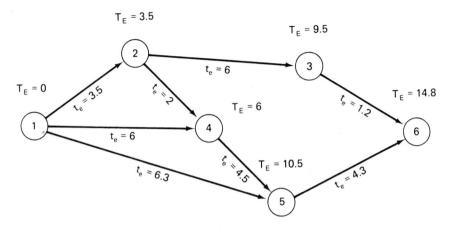

Figure 8-12   Earliest expected times for sample project.

other time prescribed on the basis of management needs. For all other events, calculation of T_L is performed using this formula:

$$T_L \text{ for Event } X =$$

$$\left(\begin{array}{c} T_L \text{ for event immediately} \\ \text{succeeding Event } X \end{array}\right) - \left(\begin{array}{c} t_e \text{ for activity immediately} \\ \text{succeeding Event } X \end{array}\right)$$

If more than one activity leads out of Event X, perform the computations for all such paths and use the *smallest* total as the T_L for Event X.

Example (using the sample project network):

		Latest Allowable Time, T_L
Event 6	The latest allowable time for the last event in the project is established as 14.8, which is the T_E for Event 6 and also the project duration.	14.8
Event 5	One path leds out of Event 5—Activity 5.6. The calculation: $\dfrac{T_L \text{ for Event 6}}{14.8} - \dfrac{t_e \text{ for Activity 5,6}}{4.3} =$ (see Figure 8-13).	10.5
Event 4	One path leads out of Event 4—Activity 4,5, which has a t_e of 4.5 weeks. $10.5 - 4.5 =$	6.0
Event 3	One path leads out of Event 3—Activity 3,6, with a t_e of 1.2 weeks. $14.8 - 1.2 =$	13.6

		Latest Allowable Time, T_L
Event 2	Two paths lead out of Event 2—(a) via Activity 2,3; and (b) via Activity 2,4:	
	(a) $\dfrac{T_L \text{ for Event 3}}{13.6} - \dfrac{t_e \text{ for Activity 2,3}}{6} =$	7.6
	(b) $\dfrac{T_L \text{ for Event 4}}{6} - \dfrac{t_e \text{ for Activity 2,4}}{2} =$	4.0
	Use the *smallest* time factor (see Figure 8-14).	4.0

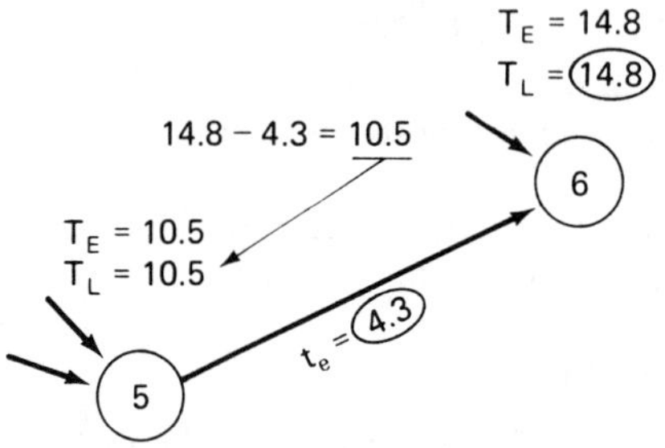

$T_E = 14.8$
$T_L = \boxed{14.8}$

$14.8 - 4.3 = 10.5$

6

$T_E = 10.5$
$T_L = 10.5$

$t_e = \boxed{4.3}$

5

Figure 8-13 Latest allowable time (T_L) for Event 5.

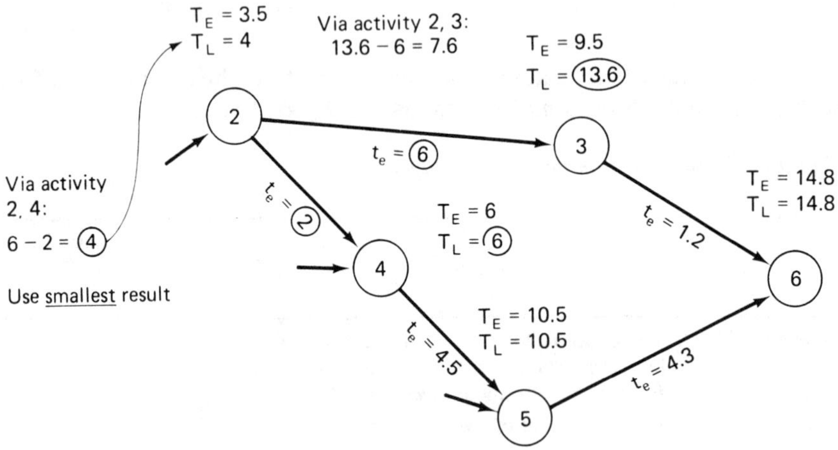

$T_E = 3.5$
$T_L = 4$

Via activity 2, 3:
$13.6 - 6 = 7.6$

$T_E = 9.5$
$T_L = \boxed{13.6}$

2

$t_e = \boxed{6}$

3

Via activity 2, 4:

$6 - 2 = \boxed{4}$

Use smallest result

$t_e = \boxed{2}$

$T_E = 6$
$T_L = \boxed{6}$

4

$t_e = 1.2$

$T_E = 14.8$
$T_L = 14.8$

6

$T_E = 10.5$
$T_L = 10.5$

$t_e = 4.5$

$t_e = 4.3$

5

Figure 8-14 Latest allowable time for Event 2.

Event 1	Three paths lead out of Event 1, via these activities: (a) 1,2; (b) 1,4; and (c) 1,5.	
	(a) $4 - 3.5 = 0.5$	
	(b) $6 - 6 = 0$	
	(c) $10.5 - 6.3 = 4.2$	
	Use the *smallest* time factor.	0

The network for the sample project, shown in Figure 8-15, includes all of the latest allowable times for events.

Slack (Event)

The *slack* of an event is the amount of time that the "completion of the event" can be delayed (if any) without advancing the completion date of the project. This "event-oriented" terminology means that, for example, if an event has 4 weeks slack, there is a 4-week interval between the two times given below:

1. The earliest time all activities immediately preceding the event will be completed according to the original time estimates (the earliest expected time, T_E).
2. The latest starting time for the most critical activity that succeeds the event (the latest allowable time, T_L).

The slack data are of great value in scheduling and controlling a project.

The formula for calculating slack is

$$T_L - T_E$$

Example: Following are the calculations of slack for the events in the sample project:

Event	$T_L{}^*$	−	$T_E{}^*$	=	Slack (Event)
1	0	−	0	=	0
2	4	−	3.5	=	0.5
3	13.6	−	9.5	=	4.1
4	6	−	6	=	0
5	10.5	−	10.5	=	0
6	14.8	−	14.8	=	0

* See Figure 8-16.

This PERT example includes events with zero slack and with positive slack (T_L is greater than T_E). In the PERT system, slack can also be negative (T_L is less than T_E). This negative slack occurs when the project duration according to the original time estimates is longer than that required to meet management needs.

To illustrate, assume the project in the sample PERT network (Figure 8-15) must be completed in 12 rather than in 14.8 weeks. All of the T_L figures in the network would have to be recomputed on the basis of the 12-week project duration:

> For Event 6, the T_L would be 12
> For Event 5, 7.7
> etc.

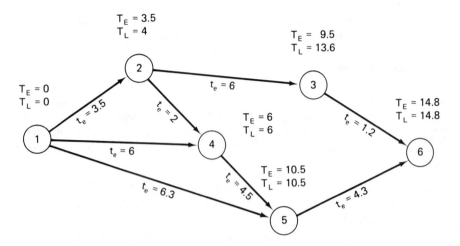

Figure 8-15   Latest allowable times for sample project.

Event 6 would then have a slack time of -2.8 ($12 - 14.8 = -2.8$); Event 5 would have -2.8 slack time ($7.7 - 10.5 = -2.8$); Event 3 a slack of 1.3 ($10.8 - 9.5 = 1.3$); and so on.

The Critical Path: The slack data indicate the critical path on the sample project (Figure 8-16). All events with "0" slack are on the critical path, which follows the route 1-4-5-6. If there is negative slack, the project may have more than one critical path. In this situation, any path with 0 or negative slack is considered a critical path. Greatest attention is given to the path with the largest amount of negative slack.

Float (Activity)

The difference between the earliest completion time and the latest completion time for an activity is the *activity float*. This is the same "float" that was calculated in the CPM approach and is used in the same manner.

The formula for calculating activity slack for Activity X is

$$\begin{pmatrix} \text{Latest completion} \\ \text{time for Activity } X \\ \text{or } T_L \text{ for the end} \\ \text{event of Activity } X \end{pmatrix} - \begin{pmatrix} \text{Earliest start time} \\ \text{for Activity } X, \text{ or } T_E \\ \text{for the beginning} \\ \text{event of Activity } X \end{pmatrix}$$

$$- \begin{pmatrix} \text{Duration} \\ \text{of} \\ \text{Activity } X, \\ \text{or } t_e \end{pmatrix} = \textit{Float}$$

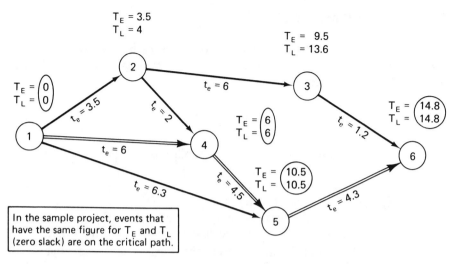

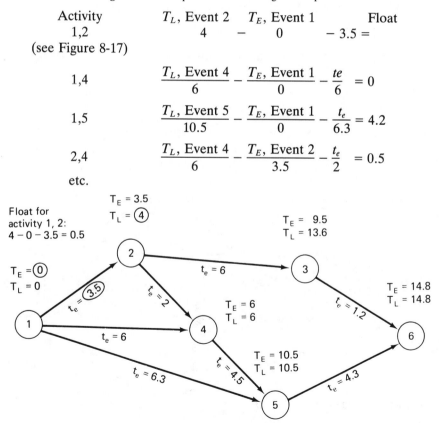

Figure 8-16 Sample network showing critical path.

Activity	T_L, Event 2	T_E, Event 1	Float
1,2	4	− 0	− 3.5 =

(see Figure 8-17)

$$1,4 \qquad \frac{T_L, \text{ Event 4}}{6} - \frac{T_E, \text{ Event 1}}{0} - \frac{te}{6} = 0$$

$$1,5 \qquad \frac{T_L, \text{ Event 5}}{10.5} - \frac{T_E, \text{ Event 1}}{0} - \frac{t_e}{6.3} = 4.2$$

$$2,4 \qquad \frac{T_L, \text{ Event 4}}{6} - \frac{T_E, \text{ Event 2}}{3.5} - \frac{t_e}{2} = 0.5$$

etc.

Figure 8-17 Sample PERT network.

CALCULATING PROBABILITY

One of the features of PERT is that it provides a method of estimating the *probability* that a schedule date can be met. Statisticians have provided the following formula for making this estimate:

$$Z = \frac{T_S - T_E}{\sigma_{T_E}}$$

where Z = measure related to the probability of meeting the scheduled date

T_S = scheduled time for the event

T_E = expected time for the event

σ_{T_E} = standard deviation—the square root of the sum of the variances of the activities used in calculating the T_E for the event

Example: A scheduled time of 13 weeks has been set for completion of the sample project. The expected time for completing the project is 14.8 weeks. Determine the probability of meeting the scheduled date.

$T_S = 13$

$T_E = 14.3$

σ_{T_E} = square root of variances for activities used in calculating T_E for Event 6 (see Figure 8-18):

Activity	Variance
1,4	0.11
4,5	1.36
5,6	1.00
	2.47

$$\sqrt{2.47} = 1.57$$

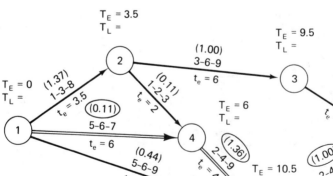

0.11 + 1.36 + 1.00 = 2.47

Figure 8-18   Sum of the variances of activities used in calculating T_E for Event 6.

Z	O	Z	O
0.0	0.5000	−3.0	0.0013
0.1	0.5398	−2.9	0.0019
0.2	0.5793	−2.8	0.0026
0.3	0.6179	−2.7	0.0035
0.4	0.6554	−2.6	0.0047
0.5	0.6915	−2.5	0.0062
0.6	0.7257	−2.4	0.0082
0.7	0.7580	−2.3	0.0107
0.8	0.7881	−2.2	0.0139
0.9	0.8159	−2.1	0.0179
1.0	0.8413	−2.0	0.0228
1.1	0.8643	−1.9	0.0287
1.2	0.8849	−1.8	0.0359
1.3	0.9032	−1.7	0.0446
1.4	0.9192	−1.6	0.0548
1.5	0.9332	−1.5	0.0668
1.6	0.9452	−1.4	0.0808
1.7	0.9554	−1.3	0.0968
1.8	0.9641	−1.2	0.1151
1.9	0.9713	−1.1	0.1357
2.0	0.9772	−1.0	0.1587
2.1	0.9821	−0.9	0.1841
2.2	0.9861	−0.8	0.2119
2.3	0.9893	−0.7	0.2420
2.4	0.9918	−0.6	0.2743
2.5	0.9938	−0.5	0.3085
2.6	0.9953	−0.4	0.3446
2.7	0.9965	−0.3	0.3821
2.8	0.9974	−0.2	0.4207
2.9	0.9981	−0.1	0.4602
		−0.0	0.5000
3.0	0.9987		

A standard table found in most mathematics and statistical handbooks.

Figure 8-19   Table of values of areas under the normal curve.

These values are substituted in the formula

$$Z = \frac{T_S - T_E}{\sigma_{T_E}}$$

$$= \frac{13 - 14.8}{\sqrt{2.47}} = \frac{-1.8}{1.57} = -1.1$$

Using the table of values of areas under the normal curve shown in Figure 8-19, for a Z of -1.1, the probability is 0.1357. This means that there is a 13.57% chance of meeting the scheduled date of 13 weeks for the project.

SUMMARY

The three time estimates in PERT—optimistic, most likely, and pessimistic—provide a basis for calculating the expected (mean) time (t_e) and variance $(\sigma_{t_e})^2$ for each activity in the project. The expected time for an activity is used in the project schedule; the variance is used in evaluating the scheduling risk involved in the expected time.

The expected times for activities are then used in calculating the earliest expected time (T_E) and the latest allowable time (T_L). With these data, the slack for each event and the float for each activity are computed. Finally, the probability of meeting the planned target date is determined.

The PERT system of planning, scheduling, and controlling a project is considered by many practitioners to be superior to the critical path method for research and development projects where many uncertainties about the schedule exist. The use of the three time estimates brings these uncertainties clearly into focus. The probability of completing the project on time is also a useful number to those versed in probability theory.

In recent practice, however, many PERT users have dropped the probability features. Several problems have been encountered with them. One is that estimators tend to be overly pessimistic when giving pessimistic times, thereby biasing project completing time toward the pessimistic side. Another is that the validity of the probability calculations themselves has been questioned.

Some installations are now using "activity-oriented" rather than "event-oriented" PERT. In the case where probabilities have been dropped and activity-oriented PERT is used, one has difficulty in seeing any differences between PERT and CPM. Since the development of the PERT/COST system, which is discussed in Chapter 9, it is customary to refer to the PERT concept described above as PERT/TIME.

The similarities between PERT and CPM are very pronounced, and it appears that the PERT originators had knowledge of CPM before they derived their system. It is interesting to note, however, that the author is not aware of any extension to PERT that considers the time/cost compression aspect, which is such a valuable extension of CPM.

9
PERT/Cost

There is a continuing need for management methods that will assist in defining the work to be performed and in developing more realistic schedule and cost estimates based on the resources planned to perform the work. These methods should also assist in determining where resources should best be applied to achieve the time, cost, and technical performance objectives, and to identify those areas developing potential delays or cost overruns in time to permit corrective action. For example, managers at each level should be able to determine:

1. Whether the current estimated time and cost for completing the entire project are realistic.
2. Whether the project is meeting the committed schedule and cost estimate and, if not, the extent of any difference.

The first step is to prepare a feasible PERT plan and schedule. Next, the PERT/COST technique requires periodic comparisons of the actual costs incurred for each activity and the actual time observed by each activity with their

The material in this chapter is presented in a condensed and modified form from the *DOD and NASA Guide, PERT/COST Systems Design,* Office of the Secretary of Defense and National Aeronautics and Space Administration.

original estimates. This comparison significantly improves cost and schedule control by establishing the cost and time status of the project and identifying any potential cost overruns and schedule slippages. Estimates of cost and time needed to complete work not yet started are also obtained in order to predict future slippages and cost overruns.

PERT/COST REPORTS

The basic information generated in the PERT/COST system can be summarized in several ways for program management reporting. The format and detail depend upon the planning and control requirements of each level of management and vary accordingly. Essentially, the reports provide the following information:

- The current project plan, schedule, and budget
- Time and cost performance to date, in relation to the plan
- Time and cost projections for completion of the project objectives

These PERT/COST reports point out potential trouble spots in the project and make it possible for managers to anticipate schedule slippages and cost overruns/underruns. Since the reports rank problem areas according to how critically these areas affect the total project, managers know where their attention is most urgently needed.

The PERT/COST reports summarize current information without distortion to each level of project management, thereby relieving a manager of the need to review detailed data from subordinate levels in order to evaluate project status. However, this detailed information is available without additional processing should any specific area require analysis.

All the PERT/COST reports are interrelated, each dealing with the same basic data, but each emphasizing a different element of the project. PERT/COST reports would usually be prepared on a monthly basis to coincide with normal accounting procedures. However, reporting can be done more frequently if required or desired.

Management Summary Report

The PERT/COST *management summary report* (Figure 9-1) shows the overall schedule and cost status of the project as a whole as well as each of the major component items. The report provides the following information:

- *The cost overrun/underrun to date,* through a comparison of the estimated costs with actual costs for the work performed.
- *The projection of cost overrun/underrun for the total project,* which is

PERT/COST
Management summary report

Program: MWS | Summary level: 4 - Controls | Report covers the period: 1 July 1961 - 31 March 1962
Project: A10 Vehicle | Contractor: | Contract | Date this

PERT/COST
Management summary report

Program: MWS | Summary level: 4 - Case | Report covers the period: 1 July 1961 - 31 March 1962
Project: A10 Vehicle | Contractor: | Contract | Date this

PERT/COST
Management summary report

Program: MWS		Summary level: 3 - Propulsion	Report covers the period: 1 July 1961 - 31 March 1962
Project: A10 Vehicle	Contractor: Missile Systems Co.	Contract number 659	Date this report: 3/31/62

Item	Cost of work: Work performed to date $: Original estimate	Actual costs	Overrun (underrun)	Totals at completion $: Contract estimate	Latest revised estimate	Projected overrun (underrun)	Schedule (*- Scheduled completion date of total item; E - Earliest completion date, L - Latest completion date: Of most critical element within item; 1961 J A S O N D, 1962 J F M A M J J A S O N D, 1963 J F M A M J)	Day	Slack status (weeks)	Remarks (S = Schedule slippage, O = Cost overrun, U = Cost underrun)
Total propulsion	850,000	1,050,000	200,000	2,500,000	2,850,000	350,000	L E *	28 31	−8.0	1. Case (S, O) 2. Controls (O)
Case	72,000	186,000	114,000	596,000	814,000	218,000	L E *	12 17	−8.0	1. Mounting (S, O)
Controls	392,000	411,000	19,000	704,000	793,000	89,000	E L *	27 05	10.0	1. Staging transducer (O)
Servo	126,000	174,000	48,000	387,000	447,000	60,000	E L *	18 02	2.0	
Nozzles	67,000	64,000	(3,000)	378,000	346,000	(32,000)	L E *	28 31	−8.0	1. Cone (S, U)
Ignition	114,000	141,000	27,000	262,000	279,000	17,000	E L *	08 12	5.0	
Auxiliary power units	79,000	74,000	(5,000)	173,000	171,000	(2,000)	E L *	31 31	4.0	
							Time now			

Figure 9-1 PERT/COST management summary report.

obtained by comparing the original cost estimate for the project with actual costs plus the estimated costs to complete the project.

- *The amount of schedule slippage,* as indicated by the difference between the established schedule for project completion and the present expected date for project completion.
- *The identification of trouble spots*—identification of those areas of the project where cost or time status requires management attention.

In the example, the manager responsible for the propulsion effort would see that:

- The Propulsion System Development Effort, scheduled for completion on January 31, 1963, is now expected to be completed on March 28, a slippage of 8 weeks.
- There is a cost overrun to date of $200,000 and a projected $350,000 overrun at project completion.
- The case is the major contributor to this cost overrun, and it is also critical in terms of completing the project on schedule.
- The nozzles, also critical to the total effort, show both a cost underrun and a time slippage, and may require more resources to get back on schedule.
- The controls effort is ahead of schedule but shows a cost overrun which might indicate that extra costs are being incurred unnecessarily.

In analyzing the status of a project, the responsible manager would examine the reports for those items where trouble is indicated. He or she would then refer to the lower-level reports as required to isolate the trouble.

Labor Loading Reports and Displays

These reports are used by management to plan the application of personnel and determine the need for overtime, additional hiring, or rescheduling of activities. The *labor loading display* (Figure 9-2) shows the overall requirements for drafting labor, while the *labor loading report* (Figure 9-3) indicates the allocation of labor-hours among the various jobs.

Referring to the labor loading display and report examples, the manager can determine that:

- The irregular loading pattern may make overtime necessary in some months, even though preceding or succeeding months have unused capacity.
- The heavy loading in June 1962 is tentatively planned largely for a job with positive slack (Charge Number 39784213), while in the same time period there is a job with negative slack (39786340) which might benefit from those resources.

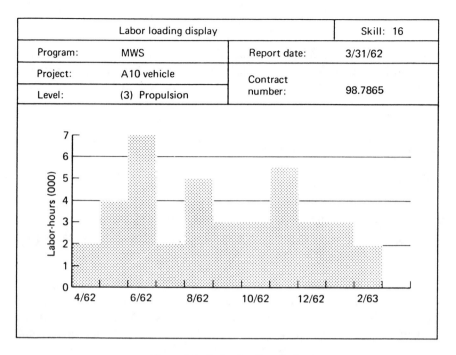

Figure 9-2 Labor loading display.

	Labor loading report			Skill: 16	
Program:	MWS		Report date:	3/31/62	
Project:	A10 Vehicle		Contract		
Level:	(3) Propulsion		number:	98.7865	
Month	Performing unit	Charge no.	Estimated labor-hours	Activity slack (weeks)
6/62	6821	39786340	1000	−4.0
	6821	39782191	2000	8.0
	5211	39784213	4000	12.0
			7000*	
7/62	6821	39782315	800	1.0
	6821	39782191	1000	8.0
	5211	39784213	200	12.0
			2000*	

Figure 9-3 Labor loading report.

- The heavy loading in June 1962 could be reduced by shifting resources within some activities from June to July or by rescheduling slack activities for July, thereby lowering the labor requirement in June.

Cost-of-Work Report

The *cost-of-work report* shows:

1. The budgeted costs to perform the work.
2. The actual costs to date.
3. The contract estimate for the work performed to date.
4. The projection of costs to project completion, based on actual costs to date and estimates-to-complete for work not yet performed.

The first two points are obtained from the financial plan. The third point is the sum of the latest revised estimates appearing on the project status report. The final point measures the progress achieved in the project. By comparing the actual costs accumulated and the contract estimate for the work performed, the manager can find out if the work is performed at a cost that is greater or less than planned.

The contract estimate for the work performed is used to identify the committed cost estimate planned for the progress achieved to date. It can be represented in this way:

$$\text{Contract estimate for all work performed to date} = \Sigma\left(\frac{A}{R} \times C\right)$$

where A = actual cost to date for a completed or in-progress work package
R = latest revised estimate for a completed or in-progress work package
C = contract cost estimate for a completed or in-progress work package

Since the project has been subdivided into work packages that must be accomplished, the PERT/COST system attempts to relate cost to progress achieved to date by comparing actual costs with the equivalent estimated costs for each work package and then summing the resultant overruns or underruns to determine the overall status of the project.

In the example (Figure 9-4), the manager can quickly determine that:

- The project is $135,000 under budget in terms of the time-phased cost plan to March 1962.
- The project is $200,000 over the cost estimate for the work completed.
- A $350,000 overrun is anticipated at the project completion.
- A schedule slippage of 8 weeks is predicted for the project.

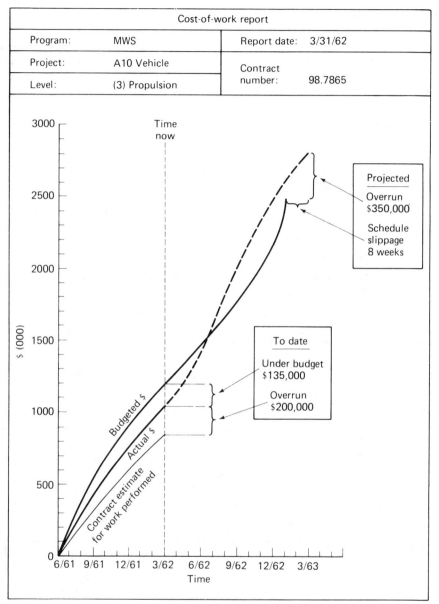

Budgeted $ = planned rate of expenditure.
Actual $ = expenditures and commitments made to date.

Figure 9-4 Cost-of-work report.

Schedule and Cost Outlook Reports

The *cost outlook report* (Figure 9-5) and the *schedule outlook report* (Figure 9-6) show the trend of successive monthly projections of the time and cost to complete the work. Each month, new projections are obtained from the project status report, and these projections provide new entries for the cost and schedule outlook reports.

For the propulsion system, the manager could determine that:

- As of the end of October 1961, the anticipated system completion was 1 week ahead of schedule, with a projected overrun of $100,000.
- As of the end of November, the outlook was for system completion in 4 weeks, ahead of schedule, but at a projected cost overrun of $200,000.
- By the end of January, both cost and schedule projections were significantly poorer.
- From the end of January to the end of March, the cost projections improved considerably, while the schedule projection deteriorated further and then showed slight improvement.

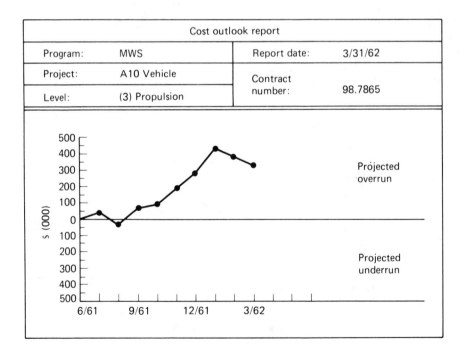

Figure 9-5 Cost outlook report.

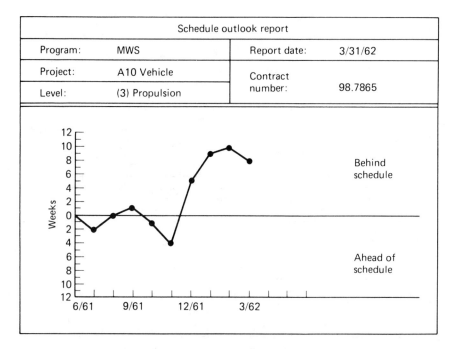

Figure 9-6   Schedule outlook report.

By relating the trend of these projections to previous management decisions, the manager can observe the effects of these decisions on cost and schedule for the project. He can determine, on a month-to-month basis, whether or not the actions taken to control schedules and costs are producing the desired results.

Milestone Reports

Selected PERT/COST network activities will represent major milestones of accomplishment toward project completion. It is quite useful to give additional project status guidance to management by reporting on these.

The milestones should be clearly identifiable in time. They should represent key network activities that are of major significance in achieving the project objectives.

The *milestone report* illustrated in Figure 9-7 shows the project manager:

- Each milestone scheduled during the current year of the program
- Action accomplished on schedule
- Action not accomplished on schedule
- Scheduled future action

Line	Milestones	Fy 1961 / Cy 1961												Fy 1962 / Cy 1962					
		J	F	M	A	M	J	J	A	S	O	N	D	J	F	M	A	M	J
1	Fuel quantity gaging				◆			⬆											
2																			
3	Engine PERT								⬆										
4																			
5	Complete R and D DEI							⬆											
6																			
7	Prototype delivered									◆⇧									
8																			
9	Letter contract award										⇧								
10																			
11	Material support plan										⇧								
12																			
13	PERT implementation											⇧⇧							
14																			
15																			
16	Deliver 1st production article																⇧⇧		
17																			
18	R and D testing ends																		⇧⇧
19																			
20																			
21																			
22																			
23																			
24																			
25																			
26																			
27																			
28																			
29																			
30																			
31																			
32																			
33																			

Current year milestone schedule

Reports control symbol
2APSC-R32

As of date
30 September 1961

Typed name and title of authenticator
FRANK J. NORTON, CAPT. USAF

Signature of authenticator
F. J. Norton

◆ Accomplished   ⬆ Past scheduled dates   ⇧ Future scheduled dates

Figure 9-7 Sample milestone report.

The milestone report illustrated in Figure 9-8 shows the project manager:

- Each milestone scheduled during the life of the program
- The scheduled completion date for each milestone
- The expected completion date for each milestone
- The latest slack calculation results for each milestone
- When and where management action is required

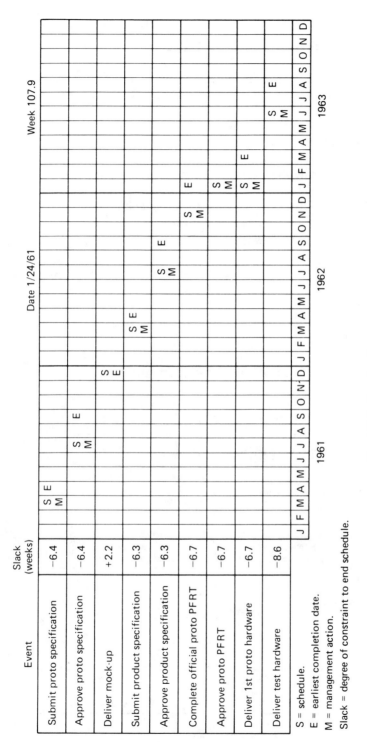

S = schedule.
E = earliest completion date.
M = management action.
Slack = degree of constraint to end schedule.

Figure 9-8   Sample milestone report.

THE PLANNING AND CONTROL CYCLE

The PERT/COST system can be represented as a cycle with two components: the planning cycle and the control cycle. This is shown in Figure 9-9.

Preparing the Project Work Breakdown Structure

By developing the work breakdown structure, we:

1. Define the project tasks to be performed and establish their relation to the project end item(s) and project objectives.
2. Establish the framework for integrated cost and schedule planning and control.
3. Establish a framework for summarizing the cost and schedule status of the project for progressively higher levels of management.

The project work breakdown structure also serves as the basis for the PERT network of activities and events.

PERT/COST system cycle

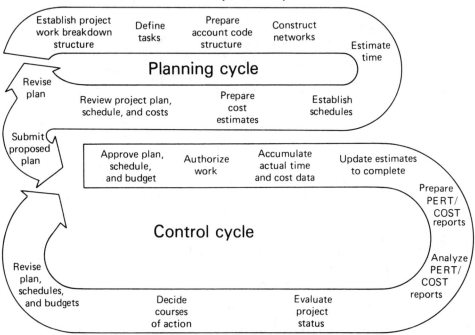

Figure 9-9   PERT/COST system cycle.

End item subdivision. The development of the work breakdown structure begins at the highest level of the project with the identification of the end items (hardware, services, equipment, facilities). The major end items are then divided into their component parts and the component parts are further subdivided into more detailed units (Figure 9-10). The subdivision continues to successively lower levels, reducing the dollar value and complexity of the units at each level, until it reaches the level where the end item subdivisions finally become manageable units for planning and control purposes. The end item subdivisions appearing at this last level are then divided into major work packages (e.g., engineering, manufacturing, testing) and responsibility is assigned to corresponding operating units.

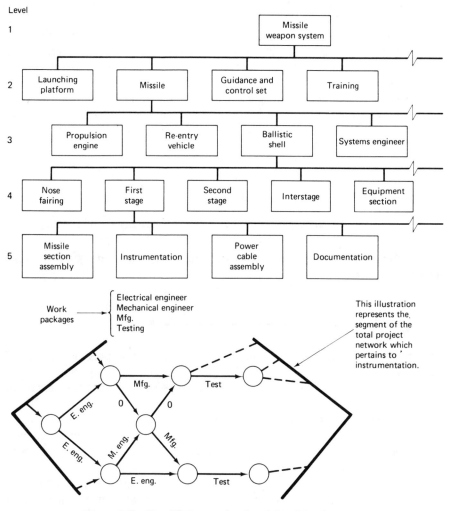

Figure 9-10 Simplified example of work breakdown structure.

The end item subdivision "Ballistic Shell Instrumentation," appearing on the lowest level of the simplified work breakdown structure, requires engineering, manufacturing, and testing work (i.e., three major work packages are identified). The engineering work package is further divided into electrical and mechanical engineering—for assignment of responsibility and for cost and resource planning, and control—because of the complexity of work to be performed. Since the responsibility for manufacturing and testing effort rests with a single unit in the contractor's organization and the dollar value is small enough, these units could be treated as separate work packages.

The number of subdivisions depends on the dollar value of the major work packages and the detail needed to control the work. Normally, the lowest-level work packages represent a value of no more than $100,000 in cost and no more than 3 months in time.

In the example, the four work packages—electrical engineering, mechanical engineering, manufacturing, and testing—as identified in the end item subdivision "Instrumentation," constitute the basic units for estimating and accumulating costs.

Network Activities: Work packages at the lowest level are usually represented by a number of activities in the PERT network, separated by events which serve as beginning or ending points for other activities in the project. For example:

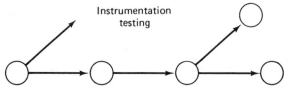

Sometimes, however, a single activity with a beginning and ending event describes a work package. For example:

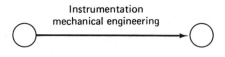

It is important to note that not all work packages will appear on the PERT network. In those cases where a work package is not directed toward a specific accomplishment of activities and events, the work need not be represented on the PERT network.

We can summarize the work breakdown structure as shown in Figure 9-11. The process does not have to occur in the exact order indicated, of course.

As the work progresses, the project work breakdown structure serves as the framework for summarizing data, so that the amount of detail presented at any level is commensurate with the decision-making requirements of management at that level. The work packages formed at the lowest level of breakdown

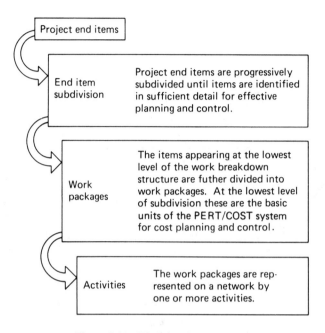

Figure 9-11   Work breakdown structure.

constitute the basic units in the PERT/COST system by which actual costs are collected.

The level of detail to which it is desirable to apply the PERT/COST system is largely a matter of judgment and varies from project to project, from one part of a project to another, and from the proposal preparation stage to the execution stage of the same project. The level of detail should be a function of factors, such as the size and complexity of the project, the degree of uncertainty, the familiarity with the work to be performed, and the time available for planning.

Preparing the Account Code Structure

The *account code structure* is the framework of numbers used for charging and summarizing the project costs. *Summary numbers* are assigned to each item subdivision on the project work breakdown structure, and *charge numbers* are assigned to each of the work packages at the lowest level of subdivision (Figure 9-12). All costs are first collected or recorded under the charge numbers assigned to the work packages. The summary numbers are then used to group or summarize costs for each end item subdivision for use by progressively higher levels of project management.

The PERT/COST system does not require costs to be estimated for each

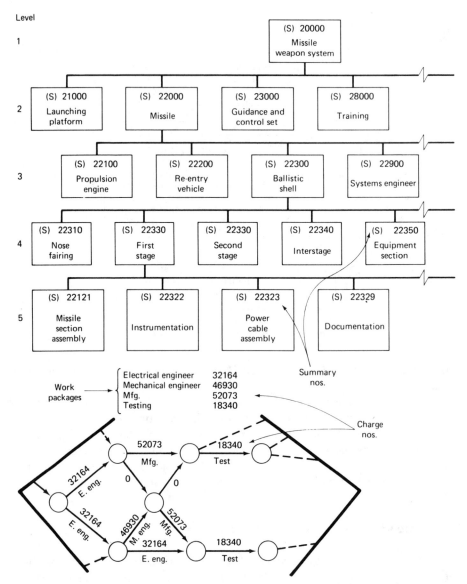

Figure 9-12 Simplified example of a work breakdown structure and account code structure.

activity on the network. This would just tend to generate expensive detail and fail to give proper weight to the relative cost significance of the various activities. Each cost activity constituting a work package is treated as a single unit for purposes of cost and resource planning and control. Again, the work package at the lowest level of subdivision is the basic unit, and all costs are collected under the same charge number.

OTHER REPORTS

Project Status Report

This report provides a summary of operations as well as detailed time and cost information. It is used to evaluate overall time and cost progress and to pinpoint areas contributing to schedule slippages and cost overruns (Figure 9-13). This information is summarized for each level of management with the summary items corresponding to the work breakdown structure.

Financial Plan and Status Report

The *financial plan and status report* assists in planning and controlling the rate of expenditure for the project (Figure 9-14). It summarizes, by future months, the total costs planned for each account. It also compares planned and actual total costs for current and past months.

UPDATING

As a project progresses, activities are added or deleted, work is completed behind or ahead of schedule, and time and cost estimates for unfinished work are revised. The PERT/COST system provides systematic *updating* by:

1. Establishing specific review dates.
2. Requiring the assignment of responsibility for estimate preparation and revision.
3. Supplying the current schedule and cost status to the individuals responsible for developing revised estimates.
4. Furnishing specific forms for transmittal of updated information.
5. Requiring reestimates when current information indicates changes in the initial estimates of personnel, material, and other resources necessary to perform the work.

PROJECT EVALUATION AND MANAGEMENT ACTION

Through a review of the output reports, the status of the project can be determined. Based on these evaluations, the project manager can take several actions to minimize costs and avoid schedule slippages. The manager can:

- Adjust the schedule of slack path activities to minimize the need for overtime or additional hiring.

Project status report

Project: Mark 93 ground support			Contract no. 98.7865			Report date: 11/30/61						
Identification				Time status					Cost status			
Charge or summary no.	Level*	Begin event no.	End event no.	Schedule elapsed time (weeks)	Date completed	Earliest completion date (S_E)	Latest completion date (S_L)	Activity slack (weeks) $(S_L - S_E)$	Actual to date $	Contract estimate $	Latest revised estimate $	Overrun (underrun) $
71831070	6	598	599	9.0	11/30/61	12/21/61	12/30/61	1.3	5,600	5,600	5,600	
71831072	6	601	602	1.0		05/03/62	03/30/62	-4.8		1,450	1,650	200
670057	5	590	602			05/03/62		-4.8	5,600	28,300	24,200	(4,100)
71831083		565	568	6.0		04/02/62	04/27/62	3.6				
		577	578	1.0		05/09/62	02/27/62	-1.7				
	6	565	578			05/09/62		-1.7	2,740	14,700	14,700	
670037	5	590	610			05/27/62		-1.7	129,000	660,000	657,900	(2,100)
670016	4	001	999			01/31/63		-4.8	889,000	3,640,000	3,665,200	25,200

*Summary level { 6. Charge level
5. Time and cost summary at major hardware level
4. Time and cost summary at subsystem level

Figure 9-13 Project status report.

Financial plan and status report					
Project: Mark 93 ground support		Contract no. 98.7865		Report date: 12/30/61	
		Dollars			
Month	Charge no.	Actual to date	Contract estimate	Latest revised estimate	Over (under) plan
Cumulative	71831017	1,200	1,100	1,200	100
Prior months	71834092	13,500	14,000	13,500	(500)
12/61	71831028	1,200	1,200	1,200	
	71834092	3,500	3,100	3,500	400
		183,700*	183,300*	183,700*	400*
01/62	71831028		1,200	1,300	100
	71834097		2,300	2,300	
	71834076		2,500	2,500	
	71831039		1,400	1,400	
		889,000**	3,640,000**	3,675,000**	35,000**

* Total by month.

** Total for project.

Figure 9-14 Financial plan and status report.

- Reallocate funds from areas of underrun to more critical areas.
- Revise the planned resources for more packages by:
  —Trading off interchangeable resources between critical and slack path activities.
  —Increasing or reducing the planned resources.
- Revise network sequence or content by:
  —Employing a greater or lesser amount of concurrence in performing activities.
  —Modifying the specifications or method of performing the work.

SUMMARY

The material covered in this session is summarized in the following under proposal preparation, negotiation, project execution, and the project evaluation.

A. PROPOSAL PREPARATION

1. Define the project objectives in terms of the project end items.
2. Prepare a project work breakdown structure showing successively further subdivisions of the project end items into functional work packages, then into activities.
3. Construct a PERT network of activities and events showing the dependency and precedence relations imposed by the technical nature of the work.
4. Prepare elapsed time estimates for each network activity based on a normal or anticipated level of resources.
5. Calculate and/or identify:
   (a) Critical path.
   (b) Slack paths.
6. Print out the basic PERT reports and compare the calculated completion dates with the directed completion dates.
7. Revise the network, if required to meet any direct completion dates, by any or all of the following methods:
   (a) Revise the network configuration to reflect a greater or lesser amount of concurrence in performing activities.
   (b) Apply a greater or lesser level of effort in performing activities.
   (c) Modify the technical approach by altering, deleting, and/or adding activities.
   (d) Recycle steps A-4 through A-6, where necessary.
   (e) Establish activity schedules.
8. Develop an account code structure conforming to the work breakdown structure in step A-2.
9. Assign charge numbers to the work packages at the lowest-level subdivision and summary numbers to all end items appearing on the project work breakdown structure.
10. Estimate, for each charge account, the labor-hours by individual or composite skills and by months required to accomplish the work in the scheduled duration as determined in step A-7e.
11. Develop an estimated dollar cost projection for each charge account. This estimate should include:
    (a) Labor costs.
    (b) Material and subcontract costs.
    (c) Special equipment, services, and other direct costs.
    (d) Indirect costs where overhead percentage rates apply.
12. Calculate:
    (a) Consolidated labor-hour requirements by months and by labor skills, showing activities with positive slack and those with negative slack occurring during the same month.

(b) A financial plan by months, including any indirect costs not accounted for in step A-11d.

13. Successively higher levels of management evaluate the plan:
    (a) Determine that personnel and other resource requirements are consistent with availability.
    (b) Determine whether idle resources and premium costs could be reduced by adjusting slack path activities.
14. Revise network plans, if required by management review, to improve labor loading and the outlook for meeting schedules:
    (a) Adjust the schedule for activities on slack paths within the limits of the earliest completion dates and the latest allowable completion dates.
    (b) Revise the planned level of resources for work packages:
        (1) Trade off interchangeable resources between critical and slack path activities, and/or
        (2) Increase or reduce the resources planned for specific activities.
    (c) Revise the network configuration to reflect a greater or lesser amount of concurrence in performing activities.
    (d) Modify the technical approach or reduce the performance requirements by altering and/or deleting, and/or adding activities.
    (e) Recycle steps A-5, A-12, and A-13, if necessary.
15. Approve schedules and budgets for the project.
16. Incorporate the schedules, budgets, and related networks into the proposed plan.

B. NEGOTIATION

1. Reaffirm or revise technical specifications and time and cost information to meet project requirements.
2. Recycle steps A-1 through A-15 to reflect any revised project requirements.
3. Contract awarded and work authorized to begin.

C. PROJECT EXECUTION

1. Charge actual labor, material, and other direct costs to account charge numbers according to normal company procedures.
2. Summarize actual cost data by the summary numbers developed in step A-9.
3. Record work progress in terms of completed activities.
4. Prepare revised time and cost estimates-to-complete for work-in-progress or work not yet started which cannot be completed within the original estimates.
5. Periodically process actual costs and schedule data and revised estimates-to-complete for preparation of PERT/COST reports.

D. PROJECT EVALUATION

1. Print out the project status report:
   (a) For completed work, print out:
      (1) Actual completion dates.
      (2) Actual costs, estimated costs, and overruns/underruns.
   (b) For work-in-process, print out:
      (1) Earliest schedule completion dates, latest schedule completion dates, and positive or negative slack.
      (2) Estimated contract costs, actual costs to date, latest revised estimates, and anticipated cost overruns/underruns.
   (c) For work not yet started, print out the latest revisions of:
      (1) Earliest schedule completion dates, latest schedule completion dates, and positive or negative slack.
      (2) Contract estimated costs, latest revised estimates, and predicted cost overruns/underruns.

2. Print out other PERT/COST reports:
   (a) Operating unit status report
   (b) Financial plan and status report
   (c) Labor loading report

3. Prepare PERT/COST management reports:
   (a) PERT/COST management summary
   (b) Labor loading report and display
   (c) Cost-of-work report
   (d) Schedule and cost outlook reports
   (e) Delivery dates for each deliverable item

4. Successively higher levels of management evaluate output reports.

5. Revise the current plan, if necessary, to avoid or minimize predicted cost overruns and/or scheduled slippages by:
   (a) Adjusting the schedule of slack path activities.
   (b) Revising the planned levels of resources for functional work packages.
      (1) Trade off interchangeable resources between critical and slack path activities.
      (2) Increase or reduce the planned resources for activities.
   (c) Revising network sequence and/or content:
      (1) Employ a greater or lesser amount of concurrence in performing the activities of the network.
      (2) Modify the technical specifications, if permissible, to allow deletion and/or addition of activities to the network.

6. Recycle steps A-5, A-12, and A-13.

7. Submit summarized cost and schedule reports.

Definitions

Activities: Project work items having specific beginning and completion points and duration times.

Activities List: A list of the project work items with accompanying descriptions which are used in preparing the network plan.

Arrow: A line with an arrowhead depicting an activity. The arrow shows direction and the passage of time.

Arrow Diagram: A graphic diagram of arrows representing component jobs and the manner in which they are connected, depicting the interrelationship among project jobs. An arrow diagram is the same as a network diagram.

Bar Chart: A chart with time activities shown as rectangular bars. The length of each bar shows duration, and the position of each bar indicates when the activity will be scheduled. It may also be called a Gantt Chart, named after its developer, Henry Gantt.

BASIC: A conversational programming language that permits the use of simple English words, abbreviations, and familiar mathematical symbols to perform logical and arithmetical operations.

Branching: An arrow diagramming error caused by two activities using the same unique designations.

Cash Flow: The income and disbursements during the span of a project.

COBOL: An acronym for COmmon Business-Oriented Language; a language formed by commonly used English nouns, verbs, and connectives, and designed specifically for application to commercial data processing problems.

Computer Hardware: The physical elements of a computer.

Computer Printout: Computer output in printed form to be used for analysis.

Computer Program: The set of instructions in computer language used by the computer to make necessary calculations.

Computer Software: The programs used to instruct the computer to perform its operations.

Cost Estimate: An estimate of cost to complete a project job based upon the resources used to accomplish the job.

Cost Slope: The rate of cost increase per unit of time duration of a work item. For example, to determine the rate of cost increase to expedite a job:

$$\text{Cost slope} = \frac{\text{Crash cost} - \text{Normal cost}}{\text{Normal duration} - \text{Crash duration}}$$

CPM (Critical Path Method): A network planning technique that is activity-oriented (utilizing arrow diagramming), used for planning and scheduling a project.

Crash Cost: The estimated cost for a job based on its crash time.

Crash Duration: The minimal time in which a job may be completed by expediting the work.

Critical Activity: A project work item on the critical path having zero float time.

Critical Path: The longest continuous path of activities through a network diagram from beginning to end of a project. The total time elapsed on the critical path, which will have zero total float, is the shortest duration of the project.

Direct Cost: The portion of the total cost that is directly related to the time in which a project item is completed. Labor costs are direct costs.

Dummy Arrow: A dashed arrow used in a network to show relationships among project items. A dummy or dummy arrow requires no time nor resources.

Duration: Activity duration is the estimated time required to complete a project job. Project duration is the total time required to complete the project as determined by the critical path. The time units may be in hours, days, or months.

Earliest Finish Time: The earliest possible time an activity can be completed without interfering with the completion of any of the preceding activities.

Earliest Start Time: The earliest possible time an activity can begin without interfering with the completion of any of the preceding activities.

Elapsed Time: The actual time required to accomplish a job in the project. Time may be measured in days, weeks, or any unit of time that is consistently considered through the project. Elapsed time used in network planning analysis includes the total length of time between the beginning and end of a project job item.

Event: A point in time shown as a circle at a junction of arrows in the network plan. It may also indicate a milestone of the completion of preceding project items. In CPM, an event is more commonly stated as a node.

Expected Time: The weighted average of the estimated optimistic, most likely, and pessimistic duration times to perform a project activity:

$$\text{Expected time} = \frac{\text{Optimistic} + 4(\text{Most likely}) + \text{Pessimistic}}{6}$$

Float: The amount of time a project job can be delayed without affecting the duration of the project. Total float is the difference between the time that is calculated to be available for a work item to be completed and the estimated duration time of that work item.

FORTRAN: An acronym for FORmula TRANslator; a computer language that uses common scientific expressions and notations in its vocabulary.

Free Float: The amount of time a designated activity can be delayed without affecting the succeeding activities.

Hammock: An activity that starts at one activity and extends in time to another activity. It is used for developing management summary reports.

i and j: Symbols designating the origin and terminal nodes, respectively, for an activity. For computer use, each activity is unique and needs to be defined by an (i,j) designation.

Independent Float: The amount of time an activity can be delayed without affecting the earliest start of the preceding activity and the latest finish of the succeeding activity.

Indirect Cost: Costs that are not associated directly with time in completing the activity. Overhead and insurance are indirect costs.

Input: Data prepared for use into a computer system.

Latest Finish Time: The latest time an activity must be completed without delaying the end of the project.

Latest Start Time: The latest time an activity can start without delaying the end of the project.

Leeway (time): Same as Float (time).

Logic Diagram: The network plan without time estimates, showing the interrelationships of project activities.

Looping: An arrow diagramming error caused by an arrow in reverse direction in the network diagram.

Management by Exception: A technique that signals problems that require the manager's attention.

Management by Objectives: A technique that defines objectives and arranges a discipline procedure to measure performance against the planned objectives.

Milestone: A major event in the project. These events may be designated as important delivery dates, major phase of building construction completion, or equipment installation completion.

Network Plan: The graphic analysis of a project, showing the plan of action through the use of a graphic diagram of arrows. *See also* Arrow Diagram.

Node: Shown graphically in the network plan as a circle depicting the beginning or end of an activity. A node represents an instantaneous point in time and occurs at the junction of arrows.

Normal Cost: The estimate of the direct cost for a project job to be performed in a normal time.

Normal Time: The estimated job time to be performed at a normal cost.

PERT (Program Evaluation and Review Technique): An event-oriented network diagramming technique used for planning and scheduling.

Precedence Diagramming: A graphic analysis of a project plan in which the nodes are the work activities (or tasks) connected by arrows. Relationships among tasks are designated as start-to-start, start-to-finish, and finish-to-finish, which eliminates the use of dummy arrows.

Project: An undertaking having a definite objective and specific beginning and completion points. The breakdown of the work items that comprise the project are set up in a logical order to achieve the objective.

Project Control: The third phase of the project management cycle that compares the actual performance of the project activities with the planned schedule and takes corrective action, where required, to avoid delays.

Project Costs Chart: A graph showing the disbursements, both plan and actual, over the length of the project.

Project Plan: The initial phase of project management, where, after setting objectives, the plan of action is developed into a logical order and portrayed in an arrow diagram.

Project Schedule: The second phase of the project management cycle detailing the time required for each job to be started and completed. The times available for starting and completing activities are predicated on the float calculations.

Resource: Money, skills, personnel, material, or equipment that may be utilized in completing a project.

Resource Allocation: Assignment of resources to each project activity.

Resource Leveling: The method of scheduling activities within their available float times so as to minimize fluctuations in day-to-day resource requirements.

Simulation: A management technique using the work plan to evaluate alternative plans to determine the best project schedule.

Slack: Used in a PERT network to show the difference between the latest allowable finish time and the earliest possible starting time of a given event. It has the same analogy as float in the CPM network.

Status Report: A communication expedient in the project control phase to inform management on the status of the project.

Time/Cost Trade-offs: A scheduling technique by which the project duration is shortened with a minimum of added costs.

Total Cost: The sum of direct and indirect costs for a work activity and/or the total project.

Total Float: The amount of time a project work item can be delayed without affecting the duration of the project. Total float time can be used in only one activity in a path.

Variance: Refers to deviations from normal costs, normal time, normal labor available, and so on.

Work Breakdown Structure (WBS): A technique that organizes resources and scope of work.

Appendix

I. PROJECT REPORT

A few of the problems relate to projects that are carried through each of the chapters and may be used to prepare a term project report. Dividing the class into groups of three to five students for the workshop sessions is timely immediately after the discussions of the illustrative problems in Chapter 2.

Each group should have similar educational interests and/or work experiences as the selection of the particular workshop project problem should be compatible with each group's background. The project problem will develop through each chapter until a complete report evolves illustrating the effective methods for planning, scheduling, and controlling a project.

Each student, upon completion of this term project report, will have excellent reference material for future application. The project report is an excellent communication in business, government, and industry with management as it includes all of the elements of project management and arranges them in such a manner as to show an effective plan of action.

The term project contents may be outlined as follows:

A. Letter of Transmittal
B. Table of Contents

C. Introduction
D. Summary
E. Project Management Report
    1. Statement(s) of the objective
    2. List of activities with descriptions
    3. Activity groupings
    4. Subdiagram
    5. Arrow diagram, including:
        a. Time estimates
        b. Earliest start times
        c. Latest finish times
        d. Critical path
    6. Total float tabulation
    7. Schedule tabulation
    8. Bar chart schedule
    9. Cost schedule
        a. Cost slope tabulation
        b. Cost expenditure schedule tabulation
        c. Indicated cost outcome
    10. Labor schedule
        a. Labor allocation (earliest start times)
            1. Tabulation
            2. Bar chart schedule with required labor
        b. Labor leveling
            1. Tabulation
            2. Revised bar chart schedule with leveled labor
        c. Labor load charts
            1. Earliest start
            2. After leveling
F. Appendix
    1. Handout material
    2. Calculations
    3. Computer printouts

As the techniques of report writing is a course in itself, we will not attempt to offer any thorough instructions on its preparation. However, writing reports becomes an important part of a project manager's responsibilities. It may be the main line of communication. The degree to which the manager can develop a communication line will provide a measure of his or her success, not only with the relationship to members of management, but with the project itself. When communication with management has been accomplished, it is a good sign that communication with all the other members associated with the project has also been realized.

The following notes on some of the key elements of a project report may be helpful to those writing a report:

Introduction.   The introduction should briefly describe the *Purpose* (a statement of the problem), *Scope* (boundary of the problem), *Limits* (money, time, design, criteria, labor, etc.), and *Background* (facts behind the project). A report may or may not include all of the above in the introduction; however, in some instances they may be combined.

A two or three paragraph introduction is usually adequate. It is advisable to review the introduction after completing the other sections of the report as the scope may have been enlarged. Refining the complete introduction is also recommended upon completion of the report.

Summary.   The summary presents the entire work on approximately one typewritten page. In recent years this page has been titled *highlights* or *executive highlights* and is made up of numbered or bullet (●) statements. Almost all of its contents will be the decisions and judgments (also known as conclusions and recommendations) based on the findings given in the report.

Place the summary (or *highlights*) immediately after the introduction as it has a better chance of being read. Management usually has a greater interest in what the report has accomplished rather than the method in which the findings were made.

Conclusions.   Conclusions are the primary reason the report is written; prepare clear, concise, and direct statements and arrange them in some order. While conclusions and recommendations may be combined, it should be recognized that each recommendation is derived from one or more of the conclusions. Therefore, each recommendation must have some support from the conclusions. Whenever possible, provide alternate courses of action, or options, and a preferred approach.

Description (project management report).   The description is the main body of the report and should represent the "meat" of the report—what was done, how it was done, and what the solutions were. The narrative portions of the body of the report are usually written in the third person—we, they, etc. Use headings and subheadings to organize the material into categories of information. (The report outline represents a good reference for headings and subheading titles.)

Appendix.   The appendix is a collection of explanatory material which cannot be given in the body of the report without disturbing its logical development. The appendix includes calculations, reference literature, supportive charts, and any other notes that do not pertain directly to the report.

Wherever possible, place detailed data in the appendix. Place detailed diagrams as near the initial reference as possible.

II. TEXT PROBLEMS

Chapter 1

1. Define the following project management principles:
   a. Network analysis
   (b) Management by objectives
   (c) Management by exception
   (d) Cost scheduling
   (e) Cost minimizing
   (f) Resource allocation
   (g) Resource leveling
   (h) Participative management/employee involvement
2. A project management cycle consists of three major phases. Identify them and provide a brief definition of each.
3. Prepare a tabulation showing advantages and disadvantages of bar charts or Gantt charts. Do the same for the network diagramming method.
4. What are the differences between the Critical Path Method (CPM) and Project Evaluation Review Technique (PERT) methods of network diagramming?

Chapter 2

1. Draw a diagram for the following:
   Project consists of five Jobs: A, B, C, D, and E.
   At project start, Jobs A and B can begin.
   Job C follows Job A (only).
   Job D follows A and B.
   Job E follows Jobs B and C.
   When Jobs D and E are finished, the project is complete.
2. Develop a network diagram for the following projects where the major project jobs have been defined.
   a. Theater planning

(1) Form organization	(8) Conduct publicity campaign
(2) Complete financing	(9) Conduct rehearsals
(3) Select play	(10) Conduct dress rehearsals
(4) Contact stars	(11) Book out-of-town shows
(5) Select director	(12) Procure scenery
(6) Select cast	(13) Conduct out-of-town shows
(7) Select set designer	(14) Prepare for opening night

   b. Changing a flat tire: Assume that there are four people in the car who can help in changing the tire.

(1) Remove flat	(3) Get lug wrench
(2) Remove lugs	(4) Get spare tire

(5) Remove hubcap
(6) Get jack
(7) Position jack
(8) Loosen lugs
(9) Stop car
(10) Replace screwdriver
(11) Jack-up car
(12) Put on spare
(13) Open trunk
(14) Replace lugs

(15) Put flat in car
(16) Get screwdriver
(17) Lower car
(18) Replace jack
(19) Close trunk
(20) Replace wrench
(21) Drive off safely
(22) Replace hubcap
(23) Tighten lugs

c. Getting Dad ready for work: Dad, Mother, and the two children help Dad to arrive at work as efficiently as possible. Listed below are a group of activities, and whether they are performed by Dad (D), Mother (M), or the children (C):

(1) Set alarm   (D)
    (night before)
(2) Awake to alarm   (D)
(3) Wash   (D)
(4) Shave   (D)
(5) Dress   (D)
(6) Set Table   (C)
(7) Fry eggs   (M)
(8) Make coffee   (M)

(9) Eat breakfast   (D)
(10) Put dishes in sink   (C)
(11) Put on hat and coat   (D)
(12) Wake wife   (D)
(13) Wake kids   (D)
(14) Start car and warm
     up   (D)
(15) Kiss wife goodbye   (D)
(16) Drive to work   (D)

3. (For Industrial Supervision students): You are required to move your production operation from the existing location to a new plant 25 miles away. Develop a plan to make the move with minimum lost production time.
   • All of the existing equipment will be reinstalled.
   • There are two similar production lines, and they will be reinstalled.

4. (For Engineering students): Your architectural engineering firm has been chosen to prepare designs and specifications for a large building project. Develop a plan for identifying the responsibilities of the civil, electrical, and mechanical engineering departments to implement this project.

5. (For Computer Systems students): As systems manager of a manufacturing firm, you are assigned the task of developing a plan for an information system. Prepare a simple diagram showing the major activities in developing such a system.

6. Prepare a sample network of 10 to 15 jobs for constructing a house. (Include such major activities as constructing foundation, framing, siding, millwork, electrical, plumbing, building and painting walls, decorating interior, and landscaping.)

Chapter 3

1. Given the network diagrams, compute the earliest start and latest finish times, and total float. Locate the critical path on the network diagrams.
   (a) Lawn project (Students to estimate durations)
       1. One-person job (Figure A-1)
       2. Two-person job (Figure A-2)
   (b) Changing a flat tire (Figure A-4)

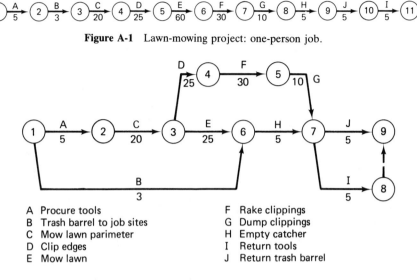

Figure A-1 Lawn-mowing project: one-person job.

A Procure tools
B Trash barrel to job sites
C Mow lawn parimeter
D Clip edges
E Mow lawn

F Rake clippings
G Dump clippings
H Empty catcher
I Return tools
J Return trash barrel

Figure A-2 Lawn-mowing project: two-person job.

(c) Getting Dad ready for work (Figure A-5)
(d) Theater planning (Figure A-3)
(e) New equipment installation (Figure A-6)

2. For each of the projects in Problem 1, complete a worksheet showing schedule information as follows:

(a) Job
(b) Description
(c) Duration
(d) Earliest start time
(e) Earliest finish time

(f) Latest start time
(g) Latest finish time
(h) Total float
(i) Free float
(j) Independent float

3. Refurnishing an office—The following activities must be accomplished to complete an office remodeling project:

Activity	Estimated Duration (days)
Procure paint	2
Procure new carpet	5
Procure new furniture	7
Remove old furniture	1
Remove old carpet	1
Scrub walls	1
Paint walls	2
Lay new carpet	1
Move in new furniture	1

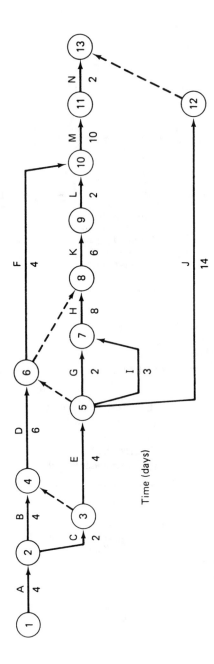

Time (days)

A 1, 2   Form organization
B 2, 4   Complete financing
C 2, 3   Select play
D 4, 6   Contract stars
E 3, 5   Select director
F 6, 10 Book out-of-town shows
G 5, 7   Select set designer

H 7, 8   Procure scenery props
I 5, 8   Select cast
J 5, 12 Conduct publicity campaign
K 8, 9   Conduct rehearsals
L 9, 10 Conduct dress rehearsals
M 10, 11 Out-of-town shows
N 11, 12 Opening night preparations

Figure A-3   Theater planning.

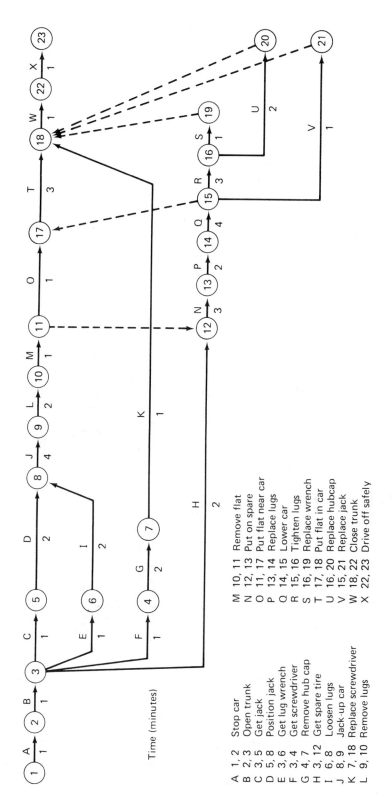

Figure A-4 Changing a flat tire; four-person job.

Time (minutes)

A 1, 2   Stop car
B 2, 3   Open trunk
C 3, 5   Get jack
D 5, 8   Position jack
E 3, 6   Get lug wrench
F 4, 7   Get screwdriver
G 4, 7   Remove hub cap
H 3, 12   Get spare tire
I 6, 8   Loosen lugs
J 8, 9   Jack-up car
K 7, 18   Replace screwdriver
L 9, 10   Remove lugs

M 10, 11 Remove flat
N 12, 13 Put on spare
O 11, 17 Put flat near car
P 13, 14 Replace lugs
Q 14, 15 Lower car
R 15, 16 Tighten lugs
S 16, 19 Replace wrench
T 17, 18 Put flat in car
U 16, 20 Replace hubcap
V 15, 21 Replace jack
W 18, 22 Close trunk
X 22, 23 Drive off safely

197

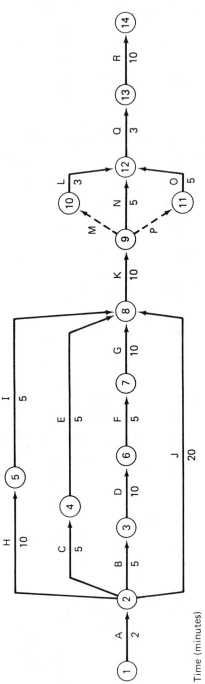

Time (minutes)

A 1, 2 Set alarm
B 2, 3 Awake to alarm
C 2, 4 Wake wife
D 3, 6 Wash
E 4, 8 Fry eggs
F 6, 7 Shave
G 7, 8 Dress
H 2, 6 Wake kids
I 6, 8 Set table

J 2, 8   Make coffee
K 8, 9   Eat breakfast
L 10, 12 Put dishes in sink
M 9, 10 Dummy
N 9, 12 Put on hat and coat
O 11, 12 Start car and warm up
P 9, 11 Dummy
Q 12, 13 Kiss wife
R 13, 14 Drive to work

Figure A-5 Getting Dad to work.

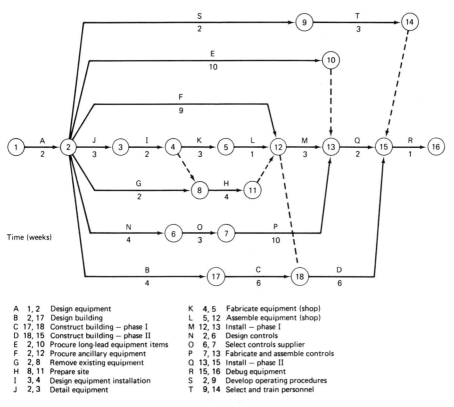

A 1, 2 Design equipment
B 2, 17 Design building
C 17, 18 Construct building — phase I
D 18, 15 Construct building — phase II
E 2, 10 Procure long-lead equipment items
F 2, 12 Procure ancillary equipment
G 2, 8 Remove existing equipment
H 8, 11 Prepare site
I 3, 4 Design equipment installation
J 2, 3 Detail equipment

K 4, 5 Fabricate equipment (shop)
L 5, 12 Assemble equipment (shop)
M 12, 13 Install — phase I
N 2, 6 Design controls
O 6, 7 Select controls supplier
P 7, 13 Fabricate and assemble controls
Q 13, 15 Install — phase II
R 15, 16 Debug equipment
S 2, 9 Develop operating procedures
T 9, 14 Select and train personnel

Figure A-6 New equipment installation.

(a) Draw an arrow diagram for this project.
(b) When can the new furniture be moved in?
(c) What is the project duration?

Chapter 4

(Optional assignments for Problems 1 through 6 are to prepare bar chart schedules using a project management software package.)

1. Draw a bar chart time schedule for the lawn-mowing project using two people. See Chapter 3, Problem 1a(2).
2. Draw a bar chart time schedule for the following projects whose schedules were developed in Chapter 3, Problem 2:
   (a) Theater planning
   (b) Changing a flat tire
   (c) Getting Dad ready for work
   (d) New equipment installation
3. Draw a bar chart time schedule for the computer installation project in Chapter 7.

4. Show the status on the bar chart time schedule of the new equipment installation project in its twelfth week, reflecting the following progress of the work activities:
- Prepare detail equipment   Behind 2 weeks
- Prepare site   Behind 4 weeks
- Fabricate equipment (in shop)   Behind 2 weeks
- Assemble equipment (in shop)   Behind 2 weeks
- All other work activities are progressing as planned.

Does present status indicate an extension of the present project duration? If so, show on bar chart.

5. Repeat Problem 4 for the computer installation project in its twentieth week.
- Procure computer   Behind 3 weeks
- Prepare site   Behind 2 weeks
- Develop program   Behind 2 weeks
- Procure forms   Behind 2 weeks
- Design forms   Behind 2 weeks
- All other work activities are progressing as planned.

6. Refer to Figure 4-2, the summary bar chart for the product introduction project. Show the status of this project on this chart in the fifteenth week.
- Design package and set up facility   Behind 3 weeks
- Order stock   Behind 2 weeks
- Plan and conduct advertising campaign   Behind 4 weeks
- Hire and train sales personnel   Behind 3 weeks

Chapter 5

(Optional assignments for Problems 1 and 2 are to prepare the cost schedules with a spreadsheet software package.)

1. Prepare a cost schedule for the new equipment installation project (Chapter 4, Problem 1d).

New Equipment Installation

Activity	Cost (000)
Design equipment	$ 2.0
Prepare equipment design detail	5.4
Prepare equipment installation drawings	1.0
Remove existing equipment	5.0
Prepare site	20.0
Design building	20.0
Construct building—phase I	150.0
Construct building—phase II	85.0
Fabricate equipment (in shop)	15.0
Procure ancillary equipment	54.0
Procure long-lead equipment items	60.0
Assemble equipment (in shop)	10.0
Install equipment—phase I	12.0
Install equipment—phase II	8.0
Develop operating procedures	2.0

New Equipment Installation

Activity	Cost (000)
Select and train personnel	3.0
Debug equipment	3.0
Develop and design automatic controls	1.6
Select controls supplier	1.2
Fabricate and assemble controls package	25.0
Total	$483.2

2. Prepare a cost schedule for the computer installation project (Chapter 4, Problem 3).

Computer Installation

Activity	Cost
Decide on computer	$ 3,200
Procure computer	100,000
Install and test computer	9,000
Determine site specifications	800
Solicit bids for site preparation	2,000
Award contract for site preparation	800
Prepare site	66,000
Select programming personnel	1,600
Select operating personnel	800
Train programming personnel	8,000
Train operating personnel	8,000
Layout computer records	2,000
Develop computer program	24,000
Test computer program	5,400
Design forms	2,000
Procure forms	2,000
Put program into operation	2,400
Total	$238,000

3. Reduce the project duration at minimum cost, one week at a time, until it is fully crashed (Figure A-7).

4. Boat repair: The network diagram in Figure A-8 shows that it will require 14 days to complete the boat repair. The owner

Cost Slope Calculations (Additional Cost/Week Reduced)

Job	Normal Time	Crash Time	Cost Slope
1,2	4	4	—
1,3*	5	3	750
1,7	15	12	800
2,4	10	9	1,500
3,4*	10	6	500

Cost Slope Calculations (Additional Cost/Week Reduced)

Job	Normal Time	Crash Time	Cost Slope
3,5	10	8	1,500
4,6*	10	8	1,500
5,6	9	7	2,000
6,7*	5	3	1,000

*Critical path.

(a) Complete the direct cost table.

Project Duration (weeks)	Normal Cost	Job(s) Crashed	Weeks Reduced	Crash Cost	Total Direct Costs (Normal and Crash)
30	$38,000			$	$38,000
29		3,4	1	500	38,500
28		6,7	1	1,000	39,500
27					
26					
25					
24					

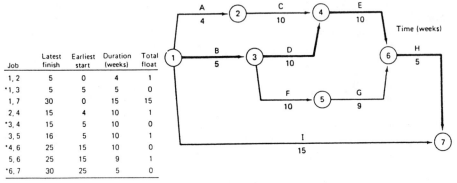

Job	Latest finish	Earliest start	Duration (weeks)	Total float
1, 2	5	0	4	1
*1, 3	5	5	5	0
1, 7	30	0	15	15
2, 4	15	4	10	1
*3, 4	15	5	10	0
3, 5	16	5	10	1
*4, 6	25	15	10	0
5, 6	25	15	9	1
*6, 7	30	25	5	0

*Critical path items.

Figure A-7

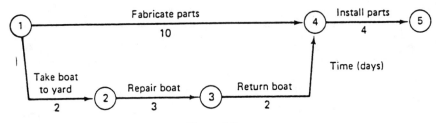

Figure A-8

of the boat had decided that he does not wish to be without the boat for 14 days and therefore is concerned with reducing the total time required for the project. He then obtains "crash" time and cost for the jobs in the project. The total normal times and costs as well as crash times and costs are tabulated below:

	Normal Data		Crash Data	
Job	Time (days)	Cost	Time (days)	Cost
1,2	2	$ 0	1	$ 20
1,4*	10	200	6	240
2,3	3	150	2	750
3,4	2	0	1	20
4,5*	4	0	4	0

*Critical path.

(a) Reduce the normal project duration at minimum cost one day at a time until it is fully crashed.

(b) Suppose that the owner decides he cannot do without the boat while the craft is being repaired; as a result, he must determine the cost of hiring a boat. After checking several boat liveries, he finds that a suitable boat can be rented for $15 per day.

(1) What is the total minimum cost for the boat repair project if it is fully crashed?

(2) What is the *total* minimum cost for each day that the project duration is reduced between the normal and crash times?

The following format may be helpful:

Project duration (days)	/	Jobs reduced	/	Normal cost	+	Extra cost	+	Cost of hiring a substitute boat	=	Total cost

5. For the new product introduction project, reduce the duration of the project at minimum cost, one week at a time, until it is fully crashed (Figure A-9).

6. Given below are the schedule and cost data for an equipment installation project:

	Normal Data		Crash Data	
Job Description	Time (weeks)	Cost (000)	Time (weeks)	Cost (000)
Design equipment	2	$ 2.0	1	$ 4.0
Prepare equipment design detail	3	5.4	2	10.8
Prepare equipment installation drawings	2	1.0	2	1.0
Remove existing equipment	2	5.0	1	12.0
Prepare site	2	20.0	2	20.0
Design building	4	20.0	3	28.0
Construct building—phase I	6	150.0	4	200.0

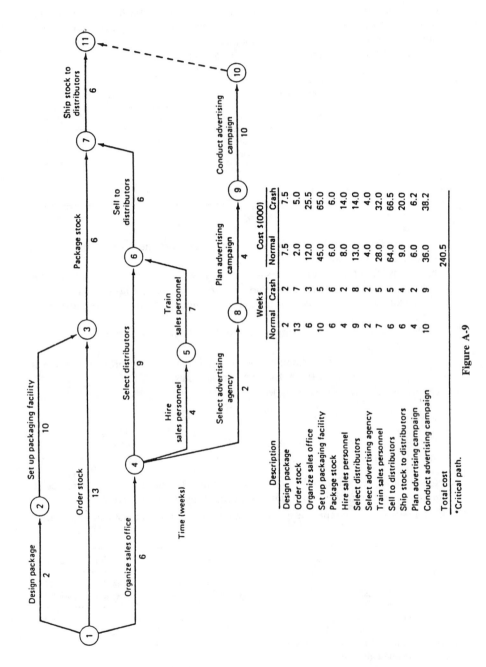

Description	Weeks		Cost $(000)	
	Normal	Crash	Normal	Crash
Design package	2	2	7.5	7.5
Order stock	13	7	2.0	5.0
Organize sales office	6	3	12.0	25.5
Set up packaging facility	10	5	45.0	65.0
Package stock	6	6	6.0	6.0
Hire sales personnel	4	2	8.0	14.0
Select distributors	9	8	13.0	14.0
Select advertising agency	2	2	4.0	4.0
Train sales personnel	7	5	28.0	32.0
Sell to distributors	6	5	64.0	66.5
Ship stock to distributors	6	4	9.0	20.0
Plan advertising campaign	4	2	6.0	6.2
Conduct advertising campaign	10	9	36.0	38.2
Total cost			240.5	

*Critical path.

Figure A-9

204

	Normal Data		Crash Data	
Job Description	Time (weeks)	Cost (000)	Time (weeks)	Cost (000)
Construct building—phase II	6	85.0	4	140.0
Fabricate equipment (in shop)	3	15.0	2	25.0
Procure ancillary equipment	9	54.0	8	70.0
Procure long-lead equipment items	10	60.0	7	80.0
Assemble equipment (in shop)	1	10.0	1	10.0
Install equipment—phase I	3	12.0	2	18.0
Install equipment—phase II	2	8.0	2	8.0
Develop operating procedures	2	2.0	1	3.5
Select and train personnel	3	3.0	2	5.0
Debug equipment	1	3.0	1	3.0
Develop and design automatic controls	4	1.6	3	2.5
Select controls supplier	3	1.2	2	2.0
Fabricate and assemble controls package	10	25.0	8	34.0

Indicated cost data, including the cost of lost production, insurance, supervision, and so on are as follows:

Total indirect cost for project duration = $100,000

Determine the minimum additional cost when reducing the project duration time by 6 weeks.

(a) Arrive at the solutions by finding the cost slope for each work item and determine indirect and direct costs for the project at 2-week intervals.

(b) Prepare a graph plotting project times on the horizontal line against project costs on the vertical side. Show direct costs, indirect costs, and total costs for a reduction of 6 weeks in project time in weekly increments.

7. Repeat Problem 6 for the computer installation project.

Total indirect cost for project duration = $20,000

	Normal Data		Crash Data	
Job Description	Time (weeks)	Cost (000)	Time (weeks)	Cost (000)
Decide on computer	4	$ 3.2	3	$ 4.4
Procure computer	25	100.0	21	120.0
Install and test computer	3	9.0	2	13.5
Determine site specifications	1	0.8	1	0.8
Solicit bids for site preparation	5	2.0	3	3.0
Award contract for site preparation	2	0.8	2	0.8
Prepare site	20	66.0	16	86.0
Select programming personnel	2	1.6	2	1.6
Select operating personnel	1	0.8	1	0.8
Train programming personnel	8	8.0	8	8.0
Train operating personnel	8	8.0	8	8.0

Job Description	Normal Data		Crash Data	
	Time (weeks)	Cost (000)	Time (weeks)	Cost (000)
Layout computer records	2	2.0	2	2.0
Develop computer program	12	24.0	8	36.0
Test computer program	3	5.4	3	5.4
Design forms	2	2.0	2	2.0
Procure forms	2	2.0	2	2.0
Put program into operation	2	2.4	1	4.0

Chapter 6

1. Given below are the labor data for the new equipment installation project:

Labor Allocation*

Job Description	Workers per Week					
	DR	D	ENG	E	P	L
Design equipment		4				
Prepare equipment design detail	6					
Prepare equipment installation drawings		2				
Remove existing equipment			1			6
Prepare site			1	2	4	
Design building	6		4			
Construct building—phase I			1	4	4	8
Construct building—phase II			1	6	6	4
Fabricate equipment (in shop)		1				
Procure ancillary equipment		2				
Procure long-lead equipment items		2				
Assemble equipment (in shop)		1				
Install equipment—phase I			1	2	4	
Install equipment—phase II			1	2	4	
Develop operating procedures		2				
Select and train personnel		2				
Debug equipment			1	2	4	
Develop and design automatic controls		1				
Select controls supplier		1				
Fabricate and assemble controls package		1				

*DR, draftsman; D, designer, ENG, engineer; E, electrician; P, pipe fitter; L, laborer.

    (a) Find the total weekly requirements for each trade for the equipment installation project. (Use the earliest start.)

    (b) Level the labor within the present project duration.

    (c) Adjust the bar chart time schedule to reflect the labor leveling effort.

2. Repeat Problem 1 for the computer installation project.

Labor Allocation

	Workers per Week	
Job Description	Analyst	Programmer
Decide on computer	2	—
Procure computer	1	—
Install and test computer	—	4
Determine site specifications	2	—
Solicit bids for site preparation	1	—
Award contract for site preparation	1	—
Prepare site	2	—
Select programming personnel	2	—
Select operating personnel	2	—
Train programming personnel	2	—
Train operating personnel	2	—
Layout computer records	—	2
Develop computer program	—	4
Test computer program	—	4
Design forms	—	2
Procure forms	—	2
Put program into operation	—	4

Chapter 7

1. Given the project planning diagrams of these projects:
   (a) Theater planning (Figure A-3)
   (b) New equipment installation (Figure A-6)

   Using a project management software package, prepare the following computer-generated reports for each of the projects above.

   (a) *Schedule report* showing earliest start and finish dates, latest start and finish dates, and float values for each activity.
   (b) *Bar chart schedule* showing initial program of project activity.

2. Given the project planning diagram of the new equipment installation (Figure A-6) and the top-level element of its work breakdown structure (WBS):

   Elements
   Site preparation
   Building
   Equipment
   Controls
   Training

   Prepare a schedule report and a bar chart schedule for (1) the project activities in a WBS sort; and (2) for each of the WBS elements.

3. Given the project planning diagram of the new equipment installation project,

prepare a milestone report showing the starting or finishing times, whichever is more appropriate, of the following key events:

(a) Start and completion of project
(b) Equipment delivery on site
(c) Building ready for initial equipment installation
(d) Controls delivery on site
(e) Start equipment debug

INDEX